I0015413

Contents

Learn to Use Visio 2016

Michelle N. Halsey

ISBN-10: 1-64004-266-0

ISBN-13: 978-1-64004-266-7

Silver City Publications & Training, L.L.C.
P.O. Box 1914
Nampa, ID 83653
https://www.silvercitypublications.com/shop/

Chapter 1 – Opening and Closing Visio

In this chapter, you will learn to open Visio. You will also explore the Visio interface, including the new Backstage view. Finally, you will learn to create a blank drawing and close Visio.

Opening Visio

Use the following procedure to start Visio.

Step 1: Press the Windows key on the keyboard to open the desktop menu.

Step 2: Select the Visio icon.

Step 3: Next, highlight the Microsoft Office program group. Select MICROSOFT VISIO 2016.

Understanding the Interface

Visio 2016 has a new interface that builds on interface from the previous version of Visio. Visio 2016 uses the Ribbon interface that was introduced in Microsoft Office 2007 applications. Each Tab in the Ribbon contains many tools for working with your drawing. To display a different set of commands, click the tab name. BUTTONS are organized into groups according to their function.

In addition to the tabs, Visio 2016 also makes use of the Quick Access Toolbar from the MS Office 2007 applications.

The File tab is a new feature that opens the Backstage View. The new Backstage View will be discussed in the next topic.

Below is the Visio interface, including the Ribbon, the Slides tab, the Slides pane, the Quick Access toolbar, and the Status bar.

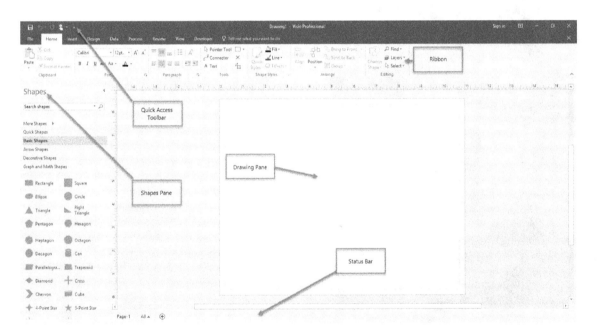

- The Shapes panel shows a thumbnail of shapes and/or stencils.

- The Drawing pane is where you can view and edit the entire drawing.

- The Status Bar provides information about your drawing and has additional tools for making changes to the view.

The Quick Access Toolbar appears at the top of the Visio window and provides you with one-click shortcuts to commonly used functions. By default, the Quick Access Toolbar contains buttons for Save, Undo and Redo.

Use the following procedure to customize the contents of the Quick Access toolbar.

Step 1: Click the arrow icon immediately to the right of the Quick Access toolbar.

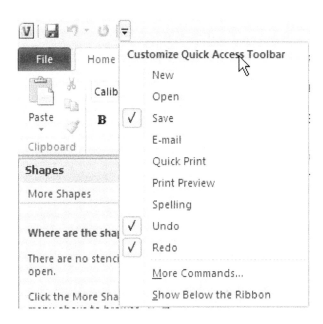

Step 2: Add an item to the Quick Access Toolbar by selecting it from the list. You can remove an item by reopening the list and selecting the item again.

If you select More Commands, Visio opens the Visio Options window.

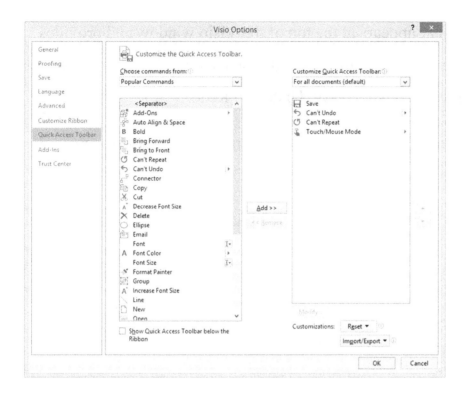

To add a command, select the item from the list on the left and select Add. Select OK when you have finished.

Using Backstage View

Select the File tab in the Ribbon to open the Backstage view. The Backstage view is where you will find the commands for creating, saving, opening, and closing files, as well as information about the file. The Backstage view includes new interfaces for printing and sharing your drawings. The Options command is also available to open a new screen for setting your Visio Options.

Use the following procedure to open the Backstage View.

Step 1: Select the File tab on the Ribbon.

Visio displays the Backstage View, open to the Info tab by default. A sample is illustrated below.

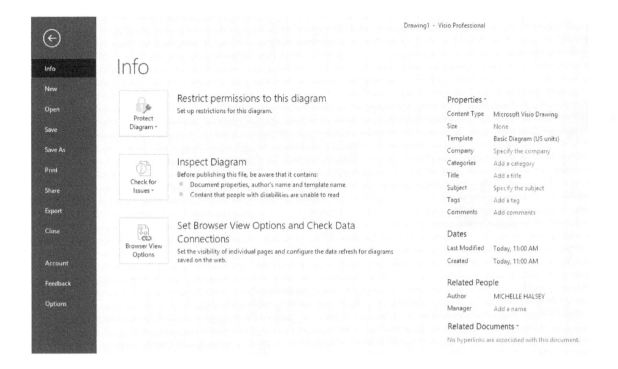

Creating a Blank Drawing

The New tab of the Backstage view provides several options for creating new drawings. The Blank Document option is at the bottom of the screen under "Other Ways to Get Started."

Use the following procedure to create a blank drawing.

Step 1: Select the File tab on the Ribbon.

Step 2: Select the New tab in the Backstage View.

Step 3: Select Blank Drawing.

Step 4: Select Create.

Closing Visio

Use the following procedure to close Visio from the Backstage View.

Step 1: Select the File tab on the Ribbon.

Step 2: Select the Close command in the Backstage View.

Chapter 2 – Working with Files

This chapter will cover some of the specific tasks you can do using the new Backstage view. First, it is important to save your work regularly to protect your work. The Backstage view allows you to open an existing file from anywhere on your computer or network. You can also easily open drawings you have recently opened. This chapter will also cover creating a drawing from a template. Finally, this chapter discusses how to close a drawing when you have finished working on it.

Saving Files

The Backstage view includes the Save and the Save As commands.

Visio will remind you to save your drawing if you attempt to close it without saving it first.

Use the following procedure to save a drawing.

Step 1: Select the File tab on the Ribbon.

Step 2: Select the Save command in the Backstage View.

If the drawing has not yet been saved, the Save As dialog box opens, so that you can name the drawing and select a location to save it. The Save As dialog box is illustrated below.

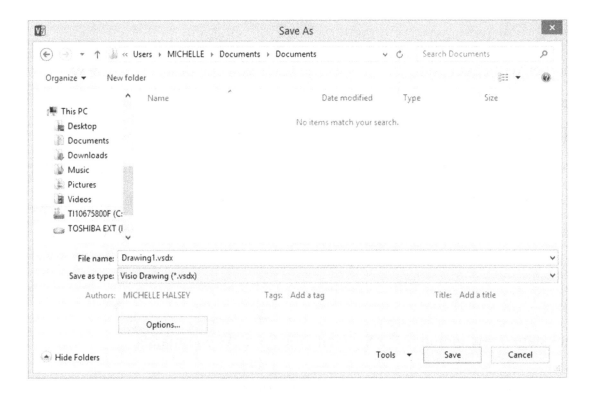

Step 3: Enter a name for the document in the File Name field.

Step 4: Use the Save in drop-down list to help you navigate to the location where you want to save the file.

Step 5: Select Save. Or you can select Cancel to close the dialog box without saving the drawing.

The Backstage view returns to the background after the save operation is complete.

Opening Files

Use the following procedure to open a drawing.

Step 1: Select the File tab on the Ribbon.

Step 2: Select the Open command in the Backstage View.

The Open dialog box opens, so that you can navigate to the location of the desired drawing and select it. The Open dialog box is illustrated below.

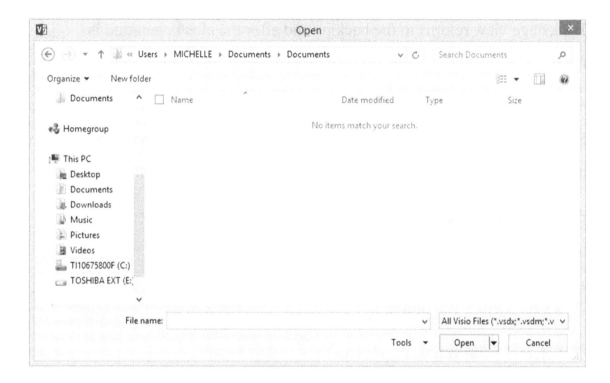

Step 3: Use the Look in drop-down list to help you navigate to the location where the file is located.

Step 4: Highlight the drawing when you find it.

Step 5: Select Open. Or you can select Cancel to close the dialog box without opening the drawing.

The Backstage view returns to the background after the open operation is complete.

Closing Files

Use the following procedure to close a drawing.

Step 1: Select the File tab on the Ribbon.

Step 2: Select the Close command in the Backstage View to close the current drawing.

The Backstage view returns to the background after the close operation is complete.

Switching Between Files

The Switch Windows tool on the View tab of the Ribbon provides a quick way to switch between drawings. You can also switch using the icon on the Status bar.

Use the following procedure to switch from one drawing to another.

Step 1: Select the View tab from the Ribbon.

Step 2: Select Switch Windows.

Step 3: Select the file you want to view.

Switch Windows tool on the Status bar.

In this chapter, you will learn how to set up your Visio screen. You have different elements to help you create your drawing, which you can show or hide as needed. This chapter will explain how to add, move, and delete a guide. It will also explain how to change the ruler settings and the grid settings.

Showing and Hiding Screen Elements

Visio's ruler, grid, guides, and connection points can be toggled on or off with the View menu. The Task panes can also be shown or hidden depending on the current need.

Use the following procedure to hide the grid lines.

Step 1: Select the View tab from the Ribbon.

Step 2: Clear the Grid box.

Open and Use Task Panes

Step 1: Select the View tab from the Ribbon.

Step 2: Select the arrow for Task Panes option to view the available task panes.

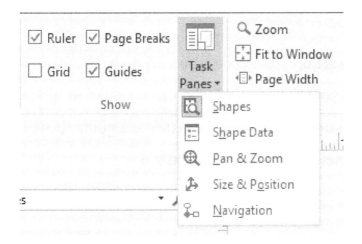

Step 3: Turn on the Task Panes you want to access. The following options are available:

Shapes – this is turned on by default

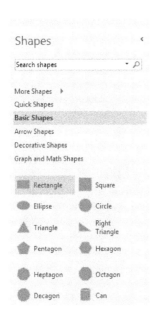

Shape Data Pane

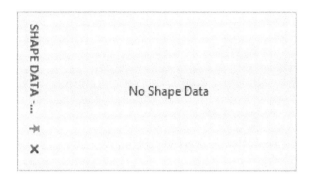

Pan & Zoom Pane

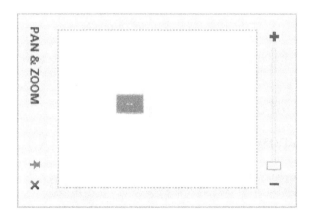

Size & Position Pane

Navigation Pane

The Task Panes

You can dock the task panes along the sides of the diagram by selecting and dragging the panes to where you want them positioned.

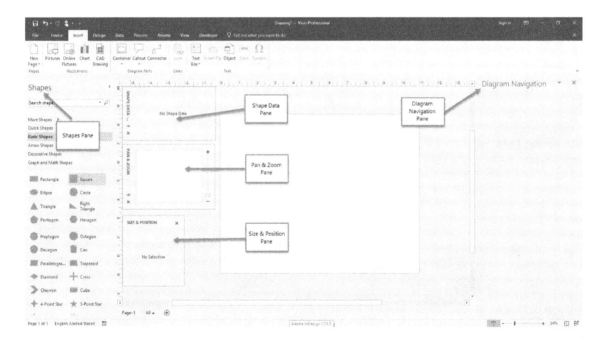

The shape data pane allows you to define properties around one of the shapes on your canvas.

Step 1: Select a shape then right-click the Shape Data pane and select Define Shape Data from the drop-down menu.

Step 2: Enter the Shape Data properties for the Shape.

Add data to selected shapes

Add data fields to your shapes and then add values to add data to shapes that are already in your drawing. The contents of the Shape Data window show the data for the selected shape or shapes only.

Add data fields to selected shapes

Step 1: Click the View tab and select Task Panes from the Show group and then click Shape Data. This will toggle display the Shape Data task pane.

Step 2: Select the shape you want to add data to.

Step 3: Right-click the Shape Data window and then select Define Shape Data.

Step 4: In the Define Shape Data dialog box, click New, and then do the following:

- Label box - type a name for the new field.
- Type list - select a type of data for the new field to contain.
- Format list - select a format for displaying the data. The format options change depending on the chosen data type.
- Text Prompt box, enter the text that you want to use as a prompt. The text appears when the mouse pointer is held over the field in the Shape Data window.

Step 5: Click OK.

Add values to your data fields

Select each shape one at a time, and type the values into the data fields in the Shape Data window or you can create a set of data fields and apply it to all selected shapes in your drawing, and to shapes that you place in the drawing later.

Create a reusable set of data fields

Step 1: Click the View tab, select Task Panes in the Show Group and then select Shape Data.

Step 2: Right-click anywhere in the Shape Data task pane, and then click Shape Data Sets.

Step 3: Click Add in the Shape Data Sets task pane.

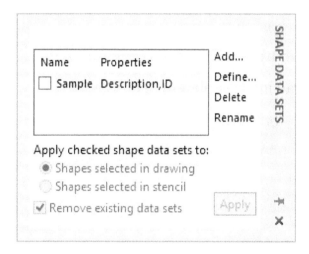

Step 4: Type a name for the data set in the Add Shape Data Set dialog box, type a name for the data set.

Step 5: Select Create a new set, and then click OK. The data set appears in the Name column of the Shape Data Sets window.

Step 6: Select your new data set, and then click Define.

Step 7: Define the fields for your shape data set in the Define Shape Data dialog box.

Step 8: Click New to create another field.

Step 9: Click Ok after all your data fields are defined.

Step 10: To apply the data set to selected shapes in your drawing, select the set in the Shape Data Sets task pane, and then click Apply.

Select the shape, and then look in the Shape Data window to see your new fields in one of the shapes that you applied it to.

Adding a Guide

Guides help you line up the shapes in your drawing. They do not appear on the final drawing when you print or save the drawing as a picture. You can add as many guides as you need, either horizontally or vertically, to help you with your drawing.

Use the following procedure to add a guide.

Step 1: Click and drag the ruler to create a guide.

Step 2: Drag it to the desired location. The guide appears as a dotted line. Release the mouse button to place the guide. The guide is still selected in the second illustration.

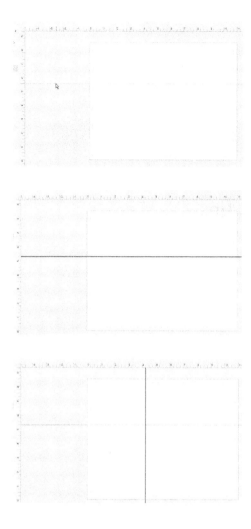

Moving or Deleting a Guide

Use the following procedure to move a guide.

Step 1: Click the guide you want to move. The small circle containing a plus sign, plus the larger blue highlighting of the guide indicates that it is selected.

Step 2: Drag it to the desired location. Release the mouse button to place the guide in the new position.

Use the following procedure to delete a guide.

Step 1: Click the guide you want to move.

Step 2: Press Delete on the keyboard to delete it.

The Rulers and Grid dialog box allows you to customize how you use the ruler and grid in the Visio canvas. You can control the ruler subdivisions or grid spacing, as well as set the origins to help with measuring objects in different positions in the drawing workspace.

Use the following procedure to open the Ruler and Grid dialog box.

Step 1: Select the square at the bottom right corner of the Show group on the View tab of the Ribbon.

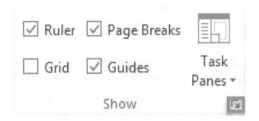

Step 2; Make the Rulers and Grid selections and then press Ok.

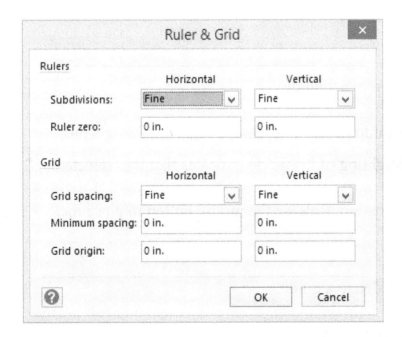

Chapter 4 – Your First Drawing

In this chapter, you will create your first drawing. Drawings consist of shapes. This chapter will cover how to find the right shape and place it on your drawing. You'll learn how to add text to shapes. You'll learn how to work with shapes, including resizing, moving, and deleting shapes. This chapter will also cover using the Tools group, which helps with refining your shapes.

Finding the Required Shape

The Shapes pane allows you to search through different categories to find the shape that you need. You can expand or collapse categories as you use them. You can use the Search Shapes tool to search for shapes within Visio, and you can even look for more shapes online.

Use the following procedure to open a stencil.

Step 1: Select More Shapes from the Shapes pane.

Step 2: Select a Category.

Step 3: Select the Shape stencil you want.

Visio displays the shapes in that stencil in the Shapes pane.

Use the following procedure to search for shapes.

Step 1: Select More Shapes from the Shapes pane.

Step 2: Select Search for Shapes.

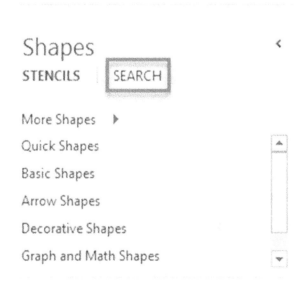

Visio displays a Search field.

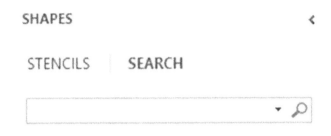

Step 3: Enter text to describe the shape you want to find.

Step 4: Press the Enter key or select the magnifying glass to begin the search.

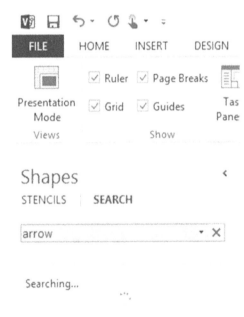

If there are many matches, Visio displays the following warning.

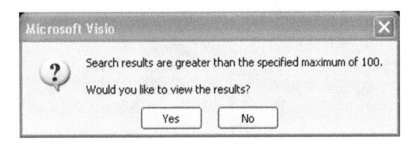

Step 5: Select Yes to continue.

Visio displays the matches in the Shapes pane. Use the scroll bar to scroll down to see all the matching shapes.

Placing the Shape in the Drawing

It is easy to drag shapes to your drawing. For greater control over positioning, use your guides or grid to help place the shapes.

Use the following procedure to add a shape to the drawing. This example shows the Glue to Guide feature.

Step 1: Click the shape you want to use.

Step 2: Drag it to the drawing canvas.

Step 3: If you want to glue the shape to a guide, position it so that the red boxes are visible.

Step 4: Release the mouse button to position the shape.

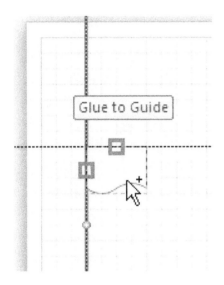

Adding Text to a Shape

Text can bring additional meaning to your diagram. To add text to a shape, simply double-click the shape.

Use the following procedure to enter text in a shape.

Step 1: Double-click the shape.

Step 2: Type your text.

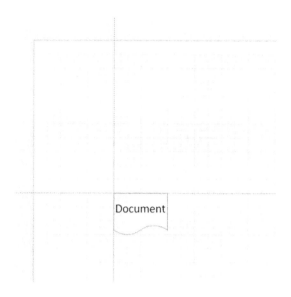

Document

Shapes can be adjusted by size and position on the drawing. You can also simply delete a shape that is not working for the drawing.

Use the following procedure to resize a shape.

Step 1: Click the shape to activate it. Visio displays handles around the shape to show that it is active.

Step 2: When you move your cursor to one of the corner handles, the cursor changes to a double-arrow. Click and drag the handle to resize the shape proportionally. Release the mouse when the shape is the size you want.

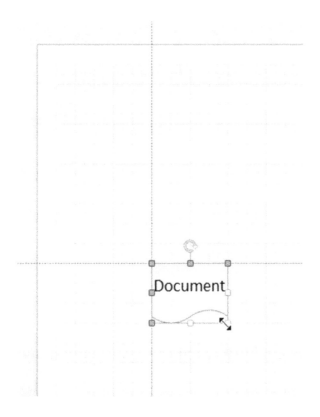

Use the following procedure to move a shape.

Step 1: Click the shape to activate it. Visio displays handles around the shape to show that it is active.

Step 2: Drag the shape. The cursor appears as a cross of double arrows.

Step 3: Release the mouse when the shape is in the new position.

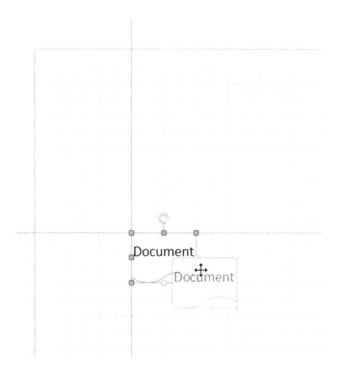

Use the following procedure to delete a shape.

Step 1: Click the shape to activate it. Visio displays handles around the shape to show that it is active.

Step 2: Press the Delete key.

Using the Tools Group

The Tools group includes a few tools to help you complete your drawing.

- The Pointer tool is used to select shapes and other objects.

- The Connector tool allows you to connect shapes.

- The Text tool allows you to select text in a shape or type free-form text on your drawing.

- The Rectangle, Ellipse, and Line tools allow you to draw basic shapes.

- The Connection Point tool allows you to add, move, or delete connection points on shapes.

The Text Block tool allows you to move, resize, or rotate the text block on a shape.

Tools group on the Home tab of the Ribbon.

Use the following procedure to connect two shapes.

Step 1: Select the Connector tool from the Home tab of the Ribbon.

Step 2: Click the shape you want to connect. The cursor changes to a red square to show that you are using a connection point.

Step 3: Drag the connector to the shape you want to connect. Release the mouse to connect the shapes.

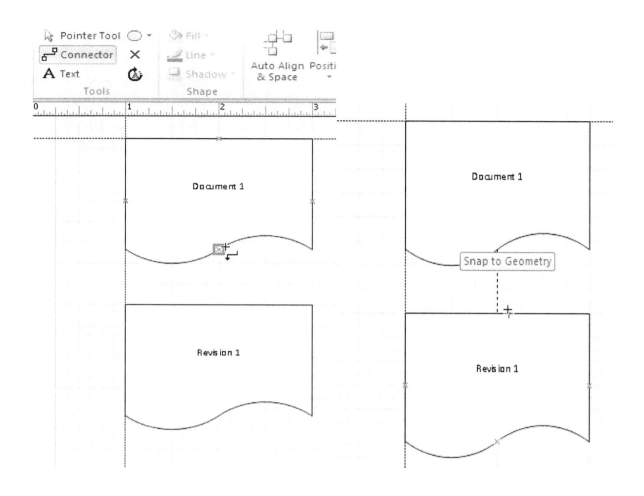

Chapter 5 – Basic Editing Tasks

The Visio 2016 editing tools make editing your drawing a breeze. This chapter covers how to cut, copy and paste text or shapes, as well as duplicating a shape. It also explains how to find and replace text, such as when you want to change a word or phrase throughout your drawing. Finally, this chapter explains how to check your text for spelling errors.

Using Cut, Copy, and Paste

Visio 2016 makes it easy to adjust drawings, including a new drawing based on a template, or a drawing started by you or another Visio user.

Before you cut or copy, select the text you want to edit by highlighting it. If you want to cut or copy a shape, just click it to select it.

The cut command deletes the selected text or shape from the current location, but allows you to move it somewhere else.

The copy command allows you to copy the selected text or shape, leaving it in the current location, but also allowing you to include it somewhere else.

The paste command allows you to include the text or shape you have cut or copied at the cursor's current location.

Click anywhere on a slide to paste text or shapes.

Use the following procedure to cut and paste text.

Step 1: Double-click the shape with the text you want to cut.

Step 2: Right click the mouse to display the context menu and select cut.

Step 3: Double-click the shape where you want to paste the text. You can also paste the text on the drawing without placing it in a shape. Just click the drawing.

Step 4: Right click the mouse to display the context menu and select Paste.

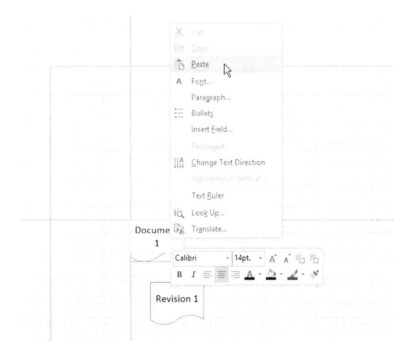

Use the following procedure to copy and paste text using the keyboard shortcuts.

Step 1: Double-click the shape with the text you want to cut and press the Control key and the C key at the same time.

Step 2: Double-click the new shape or click the drawing to paste the text outside a shape.

Step 3: Press the Control key and the V key at the same time.

Use the following procedure to copy and paste a shape.

Step 1: Click the shape you want to copy to select it. Notice the cursor changes to a cross with arrows in all directions and the shape handles are visible.

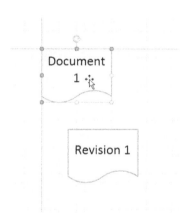

Step 2: Select Copy from the Home Ribbon, the context menu, or by using the keyboard shortcut.

Step 3: Click the drawing.

Step 4: Select Paste from the Home Ribbon, the context menu, or by using the keyboard shortcut.

Step 5: Move the shape to the new location.

Instead of dragging a shape from the Shapes pane or copying and pasting, you may want to duplicate all the features of a shape you have adjusted. Duplicating a shape makes an exact copy of the selected shape(s).

Use the following procedure to duplicate a shape.

Step 1: Click the shape to activate it. Visio displays handles around the shape to show that it is active.

Step 2: Select the Paste tool from the Home Ribbon. Select Duplicate.

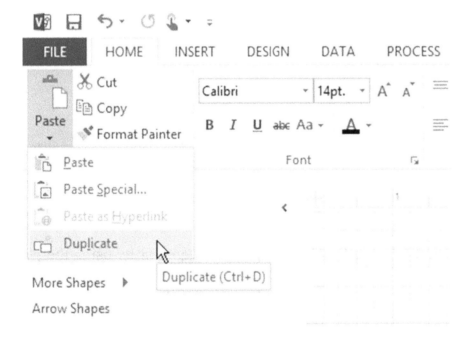

Use the following procedure to duplicate a shape using a keyboard shortcut.

Step 1: Click the shape to activate it. Visio displays handles around the shape to show that it is active.

Step 2: Press the CTRL key.

Step 3: Drag the shape. The cursor appears as an arrow with a plus sign to show that you are making a copy.

Step 4: Release the mouse when the new shape is in position.

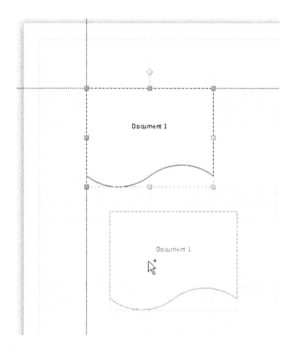

Using Undo and Redo

If you make a mistake or change your mind about your most recent task in Visio, you can undo the command. The redo command allows you to return the command results back to your drawing. The redo command also allows you to repeat tasks.

The Undo and Redo commands are so common that they appear on the Quick Access toolbar by default.

Use the following procedure to undo your most recent command.

Step 1: Select the Undo command from the Quick Access Toolbar. If there is more than one item listed, you can select more than one item to undo all selected actions.

Use the following procedure to redo the last command or repeat it.

Step 1: Select the Redo command from the Quick Access Toolbar.

Finding and Replacing Text

The Find dialog box allows you to quickly find text in your drawing. You can also search for other types of information, such as the name of a shape you used.

The Replace dialog box provides several options for finding multiple instances of text in your drawing, and replacing them, if necessary.

Use the following procedure to find text.

Step 1: Select Find from the Editing group on the Home tab of the Ribbon to open the Find dialog box. Select Find.

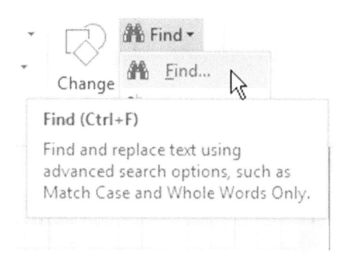

Step 2: Enter the exact text you want to find in the Find what field.

Step 3: Indicate where Visio should search for the text: the current selection, the current page, or all pages.

Step 4: Indicate whether the text is Shape text, Shape data, Shape Name or a User-defined cell by checking or clearing the boxes. Shape text is checked by default, because that is the most common selection.

Step 5: If desired, check the Match Case box to find only instances with the same capitalization.

Step 6: If desired, check the Find Whole Words only box to find the whole word. Leaving this box unchecked will find any instance of that group of letters. For example, if you search for the word box, but have the Find Whole Words Only box unchecked, Visio will find instances of box, as well as instances of "boxes" and "boxed." Check the Match character width box to only find instances of the text with the same character spacing.

Step 7: Select Find Next.

Visio highlights any matching items for you to review. Select Find Next again to find the next instance.

Visio displays the following message when it has finished searching.

Step 8: Select OK.

Step 9: Select Cancel to close the Find dialog box.

Use the following procedure to replace text.

Step 1: Select Find from the Editing group on the Home tab of the Ribbon to open the Replace dialog box. Select Replace.

Step 2: Enter the exact text you want to find in the Find what field.

Step 3: Enter the replacement text in the Replace with field.

Step 4: Indicate where Visio should search for the text: the current selection, the current page, or all pages.

Step 5: If desired, check the Match case, Match character width, and/or Find Whole Words only boxes.

Step 6: Select Find next to find the next instance of the item.

Step 7: When Visio highlights the item, select Replace to delete the "find" item and paste the "replace" item.

Step 8: Select Close when you have finished. Or select Cancel to close the dialog box without making any replacements.

Use the following procedure to replace all instances of an item.

Step 1: Select Find from the Editing group on the Home tab of the Ribbon to open the Replace dialog box. Select Replace.

Step 2: Enter the exact text you want to find in the Find what field.

Step 3: Enter the replacement text in the Replace with field.

Step 4: Indicate where Visio should search for the text: the current selection, the current page, or all pages.

Step 5: If desired, check the Match case, Match character width, and/or Find Whole Words only boxes.

Step 6: Select Replace All.

Step 7: Select Close when you have finished. Or select Cancel to close the dialog box without making any replacements.

Visio replaces all instances of the item.

Checking Your Spelling

Use the following procedure to open the Spelling dialog box.

Step 1: Select the Spelling tool from the Proofing group in the Review tab of the Ribbon.

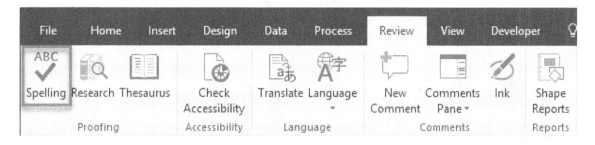

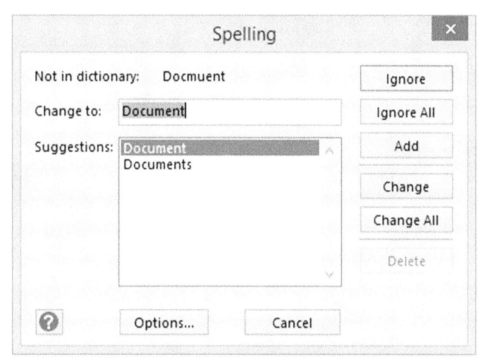

Discuss the buttons on the Spelling and Grammar dialog box.

- The Ignore button allows you to keep the word as the current spelling, but only for the current location.

- The Ignore All button allows you to ignore the misspelling for the whole drawing.

- The Add allows you to add the word to your dictionary for all Visio drawings.

- The Suggestions area lists possible changes for the misspelling. There may be many choices, just one, or no choices, based on Visio's ability to match the error to other possibilities.

- The Change To field shows the currently highlighted suggestion or you can use it to enter the correct spelling.

- The Change button allows you to change the misspelled word to the highlighted choice in the Suggestions area. You can highlight any word in the Suggestions area and select Change.

- The Change All button allows you to notify Visio to make this spelling correction any time it encounters this spelling error in this drawing.

- The Options button allows you to set the options to have Visio automatically correct certain types of errors.

Chapter 6 – Formatting Shapes

In this chapter, you'll learn how to customize your shapes. This chapter explains how to change the shape outline and fill. We'll also cover how to add shadows, how to change the line types and ends, and even how to modify the corners.

Changing the Outline

You can choose any color for your lines. The Shape group includes a gallery to choose one of the following for your line color:

Automatic – Makes the line black.

Theme Colors – Includes a palette of colors based on the drawing's theme.

Standard Colors – Includes a palette of 10 standard colors.

More Colors – Opens the Colors dialog box to choose from more colors or to enter the values for a precise color.

You can also choose from several standard line widths for your lines or shape outlines.

Use the following procedure to select a color for their lines from the gallery.

Step 1: Select the shape you want to change.

Step 2: Select the arrow next to the Line tool in the Shape Styles menu on the Home Ribbon to display the gallery. Or select the same tool from the context menu (appears when you right click a shape).

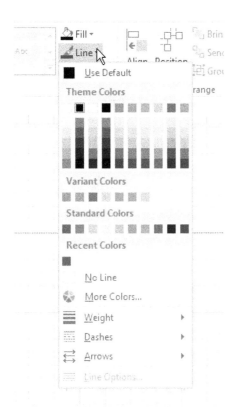

Step 3: Select the color to change the line color.

Use the following procedure to open the Colors dialog box.

Step 1: Select the shape you want to change.

Step 2: Select the arrow next to the Line tool on the Home Ribbon to display the gallery. Or select the same tool from the context menu (appears when you right click a shape).

Step 3: Select More Colors to open the Colors dialog box.

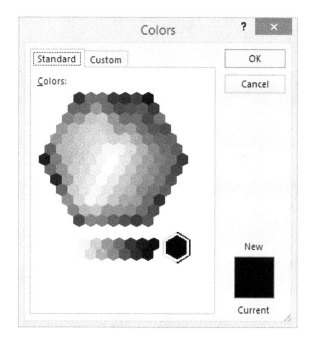

In the Standard Colors dialog box, simply click the color and select OK to use that color.

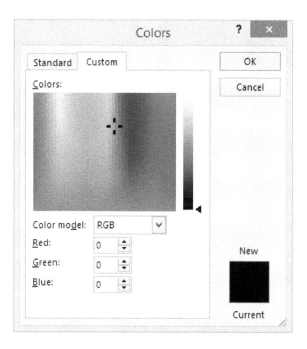

In the Custom Colors dialog box, you can click the color, or you can enter the red, green, and blue values to get a precise color. When you have the color you want, select OK.

Changing the Fill

You can choose the same or a different color for the fill in your shape. The Fill dialog box allows you to select the details for your shape fill, including various transparency levels for patterns. The dialog allows you to preview your changes before applying them.

Use the following procedure to open the Fill dialog box.

Step 1: Select the shape you want to change.

Step 2: Select the arrow next to the Fill tool on the Home Ribbon to display the gallery. Or select the same tool from the context menu (appears when you right click a shape).

Step 3: Select Fill Options from the Home Ribbon.

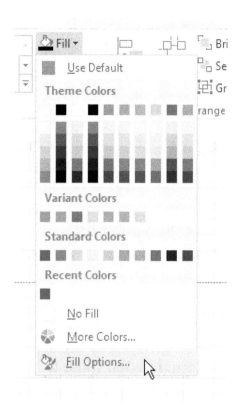

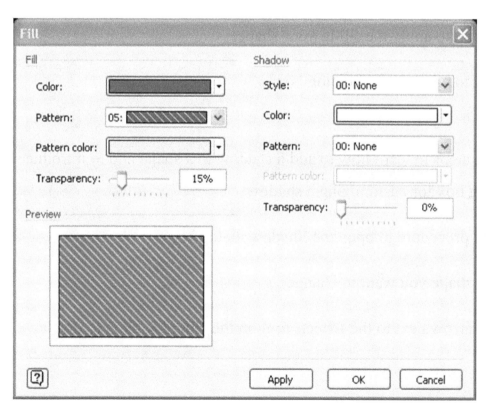

Use the following procedure to add a pattern to a shape.

Step 1: Select the shape you want to change.

Step 2: Select the arrow next to the Fill tool on the Home Ribbon to display the gallery. Or select the same tool from the context menu (appears when you right click a shape).

Step 3: Select Fill Options from the Home Ribbon.

Step 4: Select the color for the background from the Color drop-down list.

Step 5: Select the style of pattern from the Pattern drop-down list.

Step 6: Select the color of the pattern from the Pattern Color drop-down list.

Step 7: Select the Transparency for the fill color and pattern. Use the slider to select a value from 0% to 100%.

Step 8: Select Apply to apply the fill to your shape.

Step 9: Select OK to close the Fill dialog box.

Adding Shadows

Use the following steps to learn how to add a shadow to a shape and to introduce the Shadow dialog box for customizing a shadow.

Use the following procedure to open the Shadow dialog box.

Step 1: Select the shape you want to change.

Step 2: Select the arrow next to the Effects tool on the Home Ribbon to display the gallery.

Step 3: Select Shadow Options from the Effects tool options.

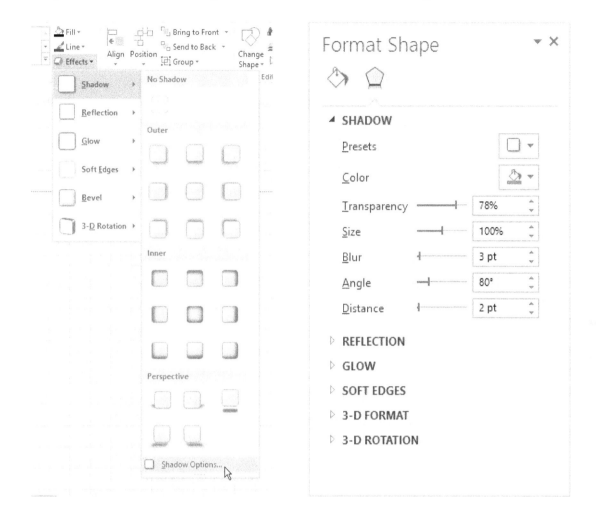

You can use a variety of line types (such as dashes) and end types (such as arrows) on your lines.

Use the following procedure to change line types.

Step 1: Select the shape you want to change.

Step 2: Select the arrow next to the Line tool on the Home Ribbon. Or select the same tool from the context menu (appears when you right click a shape).

Step 3: Select Dashes.

Step 4: Select the type of line you want to use.

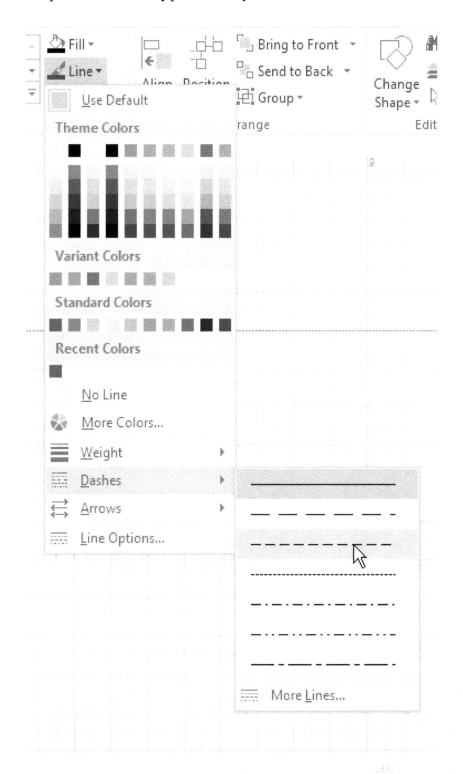

Use the following procedure to change the line end.

Step 1: Select the shape you want to change.

60

Step 2: Select the arrow next to the Line tool on the Home Ribbon. Or select the same tool from the context menu (appears when you right click a shape).

Step 3: Select Arrows.

Step 4: Select the type of end you want to use.

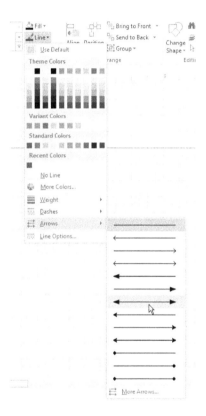

Modifying Corners

The Line dialog box allows you to change a few details for your shape lines, including modifying the corners. You can select from a gallery of different corners or customize the rounding by measurement.

Use the following procedure to apply round corners.

Step 1: Select the shape you want to change.

Step 2: Select the arrow next to the Line tool on the Home Ribbon. Or select the same tool from the context menu (appears when you right click a shape).

Step 3: Select Weight, Dashes or Arrows.

Step 4: Select More Lines or More Arrows.

Step 5: Select the type of corner that you would like to use in the Round corners area.

Step 6: Select Apply to apply the changes to your shape.

Step 7: Select OK to close the Line dialog box.

Visio 2016 allows you to enhance your text in many ways. In this chapter, we'll discuss the different types of formatting, as well as cover the most basic types of formatting for your text. This includes the font face, size, and color, as well as adding effects to the text. We'll also discuss how to use the Format Text dialog box.

Changing Font Face and Size

You can easily change the font face to any font installed on your computer. You can use the Font group on the Home Ribbon, or you can use the context menu that appears when you select text and right-click the mouse.

The Font face list includes the theme fonts first, and then the most recently used fonts, then the other fonts installed on your system in alphabetical order.

The font size is measured in points, which is unit of measure in typography.

Use the following procedure to change the font face and size using the Home Ribbon tools.

Step 1: Select the text you want to change. Or you can select the shape.

Step 2: Select the arrow next to the current font name to display the list of available fonts.

Step 3: Use the scroll bar or the down arrow to scroll down the list of fonts.

Step 4: Select the desired font to change the font of text.

Step 1: With the text still selected, select the arrow next to the current font size to see a list of common font sizes.

Step 2: Use the scroll bar or the down arrow key to scroll to the size you want and select it. You can also highlight the current font size and type in a new number to indicate the font size you want.

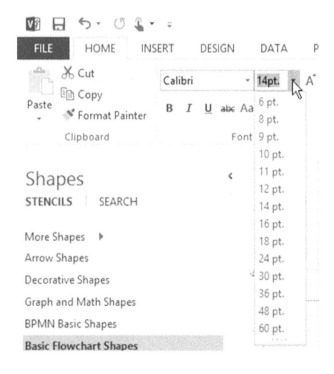

Use the following procedure to view the font context list that appears when you select text.

Step 1: Select the text you want to change.

Step 2: Right-click to display the context menu appears.

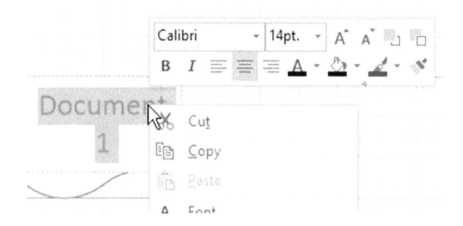

Step 3: Select the new font face or font size just as you would on the Home Ribbon.

Changing the Font Color

You can choose any color for your text. The font group includes a gallery to choose one of the following for your font color:

Theme Colors – Includes a palette of colors based on the document's theme.

Standard Colors – Includes a palette of 10 standard colors.

More Colors – Opens the Colors dialog box to choose from more colors or to enter the values for a precise color.

Use the following procedure to select a color for their fonts from the gallery.

Step 1: Select the text you want to change.

Step 2: Select the arrow next to the Font Color tool on the Home Ribbon to display the gallery. Or select the same tool from the context menu (appears when you right click).

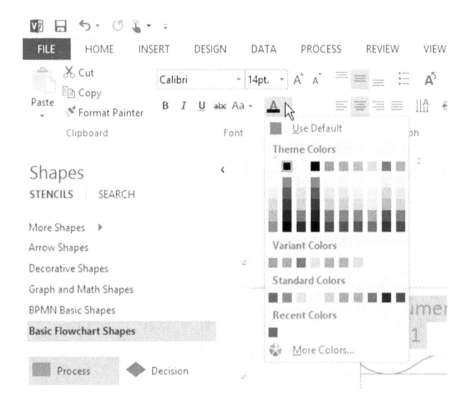

Step 3: Select the color to change the font color.

Adding Effects

You can choose several effects for your text. The font group on the Home Ribbon and the font context menu allow you to easily change the font to:

- Bold
- Italic
- Underline
- Strikethrough

Use the following procedure to add text effects.

Step 1: Select the text you want to change.

Step 2: Select the effects tool you want to use from the Home Ribbon. Bold and Italic are also available from the context menu.

Using the Format Text Dialog

The Format Text dialog box allows you to control several aspects of font formatting at one time. It also allows you to set the character spacing.

Use the following procedure to open the Format Text dialog box.

Step 1: Select the text you want to format.

Step 2: Select the square at the bottom right corner of the Font group in the Home Ribbon.

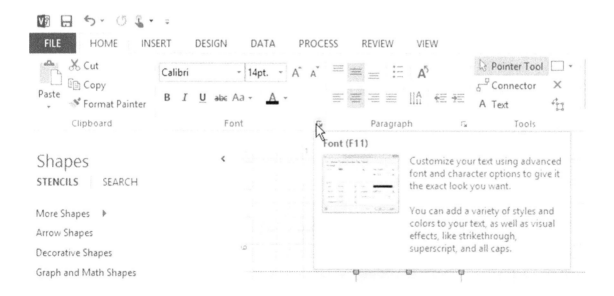

Format Text dialog box Font tab.

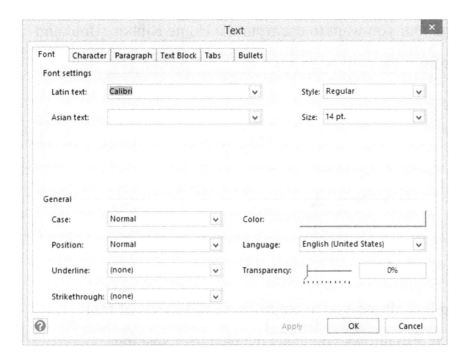

Format Text dialog box Character Spacing tab.

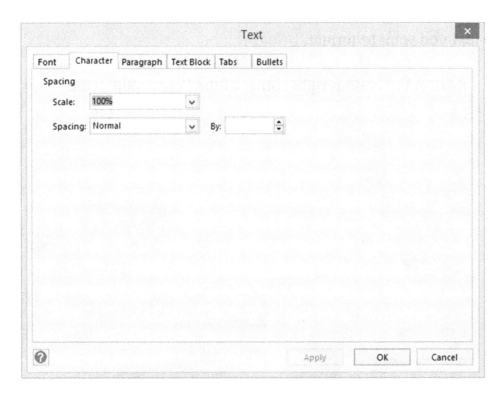

68

Chapter 8 – Formatting Blocks of Text

In Visio, text resides in blocks, whether it is part of a shape or not. In this chapter, you'll learn how to work with those text blocks. This chapter explains how to set the alignment and change the indents and paragraph spacing. This chapter also explains how to add bullets and numbering.

Setting the Alignment

You can align your text to the left, to the right, or in the center. You can also justify the text. You can also align your text to the top, to the middle, or to the bottom of the text block.

Use the following procedure to adjust the alignment for the paragraph.

Step 1: Click the text block or shape you want to adjust (the text does not have to be selected).

Step 2: Select the desired alignment tool from the Home Ribbon. You can also select multiple shapes.

Changing the Indent

You can easily indent your text, or remove an indentation, using the tools on the Home Ribbon.

Use the following procedure to adjust the indent for text.

Step 1: Click the text block or shape you want to adjust (the text does not have to be selected).

Step 2: Select the desired indent tool from the Home Ribbon. You can also select multiple shapes.

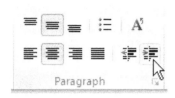

The Paragraph tab of the Format Text dialog box allows you to add space before or after a paragraph. It also allows you to adjust the line spacing within a paragraph.

Use the following procedure to open the Paragraph tab of the Format Text dialog box and adjust the space above, space below, and line spacing options.

Step 1: Click the text block or shape you want to adjust (the text does not have to be selected).

Step 2: Select the square at the bottom right corner of the Paragraph group in the Home Ribbon.

Step 3: Select the Paragraph tab.

Step 4: You can use the up and down arrows to adjust the Spacing Before and after the paragraph. The arrows adjust the points in typographical increments. You can also enter any number in the Before and After fields to adjust the spacing more precisely.

Step 6: The Line field allows you to select set the line spacing as a percentage. Single spacing would be 100%.

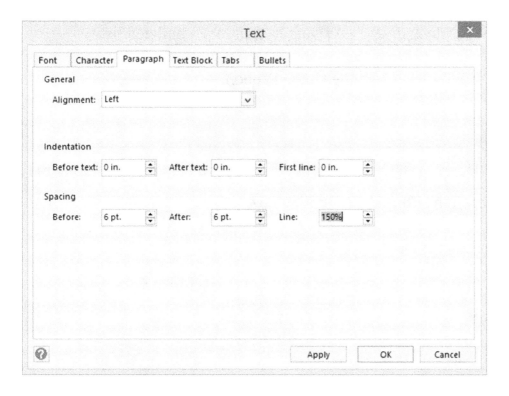

Step 7: Select Apply to apply the Paragraph settings to your text block.

Step 8: Select OK to close the Text dialog box.

Adding Bullets

The Home Ribbon includes a tool on the Paragraph group to quickly create a bulleted list. You can also use the Format Text dialog box to use custom bullets.

Use the following procedure to create a simple bulleted list.

Step 1: Select the text block or text you want to turn into a bulleted list.

Step 2: Select the Bullets tool from the Home Ribbon.

Bullet Library

Step 1: Select the square at the bottom right corner of the Paragraph group in the Home Ribbon to open the Format Text dialog box.

Step 2: Select the Bullets tab.

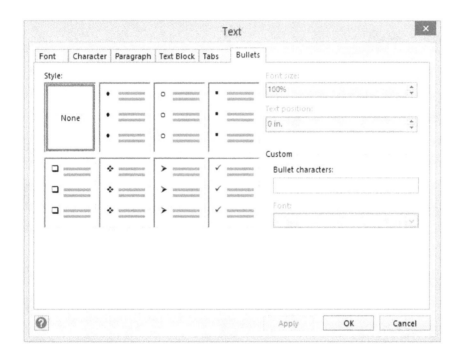

Use the following procedure to create a custom bullet.

Step 1: Select the square at the bottom right corner of the Paragraph group in the Home Ribbon to open the Format Text dialog box.

Step 2: Select the Bullets tab.

Step 3: Select the Font size for your bullet from the drop-down list.

Step 4: Enter a text position measurement or use the up and down arrows to select it.

Step 5: Enter the custom bullet character in the Bullet Characters field.

Step 6: Select the font for the bullet character from the drop-down list.

Step 7: Select Apply.

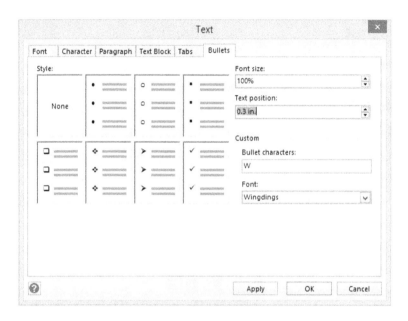

Step 8: Select OK to close the dialog box.

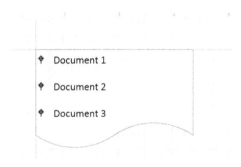

Rotating Text

In addition to aligning text, you can rotate your text blocks. The Rotate tool on the Home Ribbon rotates your entire text block counter-clockwise 90° at a time.

You can also use the Text Block tool in the Tools group to free rotate your text block to any position.

Use the following procedure to rotate text using the Rotate tool.

Step 1. Click the text block or shape you want to adjust (the text does not have to be selected).

Step 2: Select the Rotate Text tool.

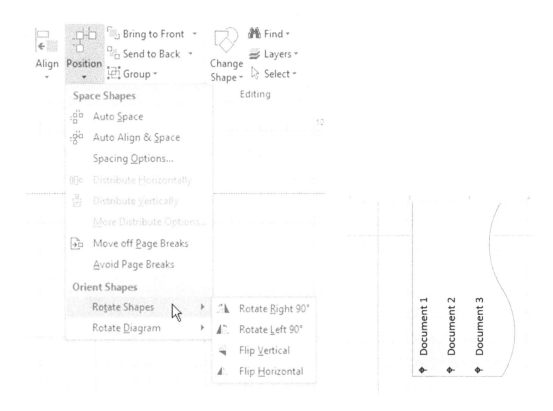

Use the following procedure to rotate text using the Text Block tool.

Step 1: Click the text block or shape you want to adjust (the text does not have to be selected).

Step 2: Select the Rotate Text tool.

Step 3: Click the top handle and drag until the text block is rotated as desired. Release the mouse to position the text block.

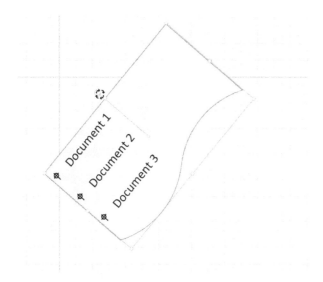

Chapter 9 – Formatting Your Drawing

This chapter explains how to format your drawing. You'll learn how to select multiple objects. This chapter covers how to use the format painter, styles, and themes to further enhance your drawing. You'll also learn how to center your drawing and change your layout.

Selecting Multiple Objects

You can resize, move, or format multiple shapes at once to save time and create a more consistent look to your drawing.

Use the following procedure to select multiple objects.

Step 1: Make sure you are using the Pointer tool.

Step 2: Draw a square around the shapes you want to select. You can also hold down the SHIFT key or CTRL key while clicking multiple shapes.

Visio highlights the selected shapes with a border around the group. Note that in the following example, the top right shape is NOT selected.

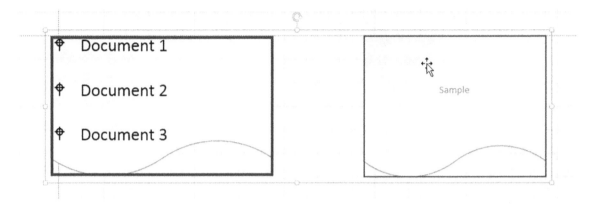

The Format Painter allows you to quickly apply the same formatting from one shape or text block.

Use the following procedure to use the Format Painter.

Step 1: Select the shape or text that has been formatted with the formatting properties that you want to copy.

Step 2: Select the Format Painter tool.

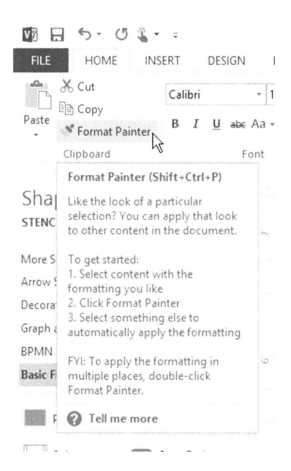

The cursor changes to a Format Painter cursor, as illustrated below.

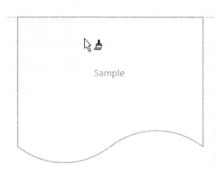

Step 3: Select the shape or text you want to format with the same properties.

The cursor returns to normal after applying the formatting properties once. You can always repeat the process to format more text or other shapes with the same properties.

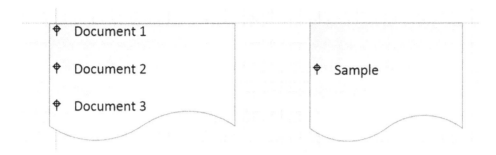

Applying a Theme

Themes control the look and feel of your entire drawing, including the colors, fonts, and shape styles. The drawing previews the themes as you hover over each option in the Themes gallery.

Use the following procedure to change the theme.

Step 1: Select the Design tab on the Ribbon.

Step 2: Select the Themes tool from the Design Ribbon to see the options.

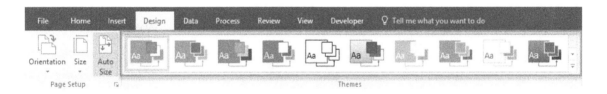

Step 3: Select a Theme from the list.

Using Backgrounds

Backgrounds are another way to customize the look of your drawing. Visio 2016 includes several background styles to quickly change the look of your drawing. You select the background style and the color separately.

Use the following procedure to change the background.

Step 1: Select the Design tab on the Ribbon.

Step 2: Select Backgrounds.

Step 3: Select the background you want to use.

Step 4: Select Backgrounds again. Select Background Color.

Step 5: Select the desired color from the color gallery.

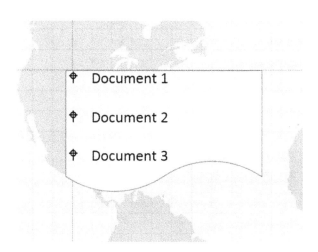

The Borders and Titles gallery provides another option for giving your drawing a professional polish. The Borders and Title option places a text block to use as a title for the drawing.

Use the following procedure to add a border and title.

Step 1: Select the Design tab on the Ribbon.

Step 2: Select Borders & Titles.

Step 3: Select the Border and Title layout you want to use.

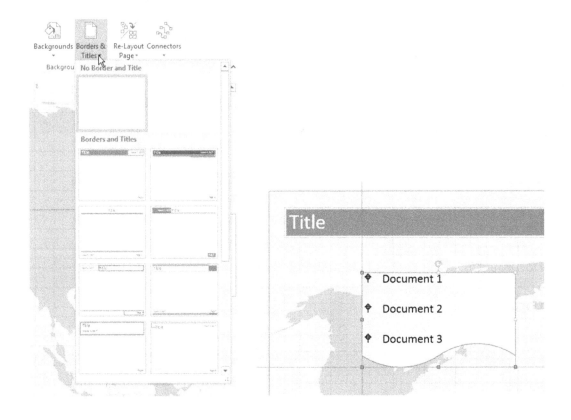

Use the following procedure to change the default text for the border and title.

Step 1: At the bottom of the drawing, there are tabs for the different pages. Click the VBackground-1 page created when you added the border and title.

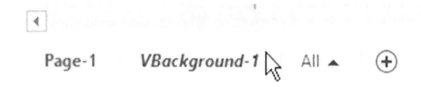

Visio displays the background layer of the drawing.

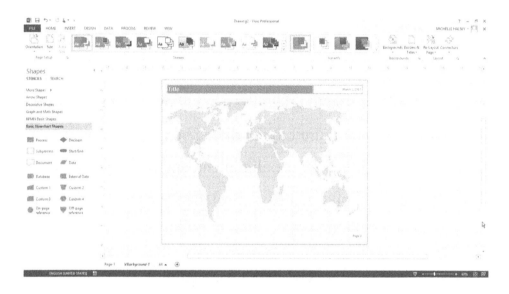

Step 2: Now you can click each Border and Title element and replace or format the text, just as with any other shape.

Step 3: Select the Page 1 tab at the bottom to return to your drawing.

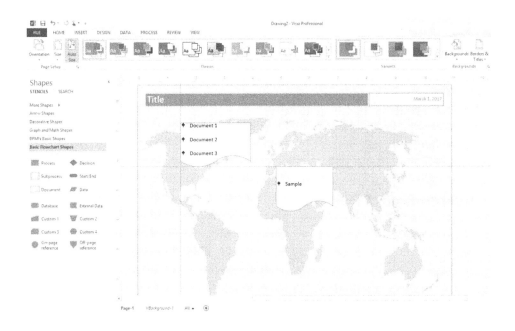

Changing Your Layout

The Re-Layout Page gallery allows you to select from many options to reposition the shapes on your drawing. The Configure Layout dialog box provides more precise control over the style, the direction, the alignment, and the spacing concerning your shape placement. You can also control the style and appearance of connectors on this dialog box.

Use the following procedure to view the Re-Layout page gallery.

Step 1: Select the Design tab on the Ribbon.

Step 2: Select Re-Layout Page.

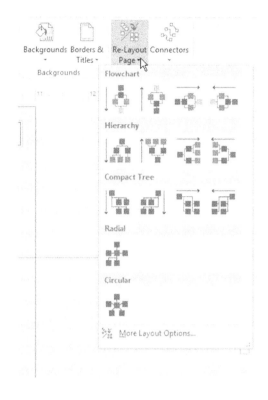

Use the following procedure to view the Configure Layout dialog box.

Step 1: Select the Design tab on the Ribbon.

Step 2: Select Re-Layout Page.

Step 3: Select More Layout Options.

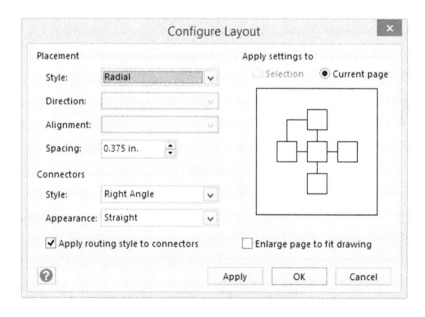

Your drawing is almost ready. In this chapter, you'll learn how to validate your drawing and add a legend. You'll also learn how to prepare the drawing for sharing. This chapter explains how to use the Page Setup group. It also explains how to save the drawing as a picture, print the drawing, and email a drawing.

Validating Your Drawing

Validation help ensure that your drawing meets general diagramming or company-specific best practices and/or requirements. You select a set of rules to use for the validation process. Once you have instructed Visio to check the diagram, it displays the Issues window to explain any deviations from the selected set of rules. You can use the Issues window to quickly find the problem shapes. You can even ignore an issue if it does not apply.

Use the following procedure to view the Check Drawing options.

Step 1: Select the Process tab.

Step 2: Indicate the rule set by selecting the arrow next to Check Diagram. Select Rules to Check. Select the rule for that diagram. Or you can select the arrow next to Check Diagram and select Import Rules from. Then select the rules from another open Visio file.

Step 3: Check the diagram by selecting Check Diagram.

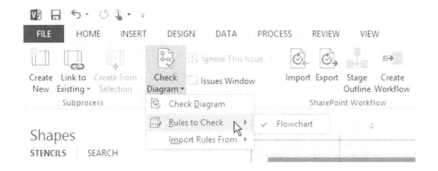

Use the following procedure to view the Issues Window.

Step 1: Make sure that the Issues Window box on the Process tab is checked.

Step 2: After you check a drawing, view the Issues window.

Step 3: You can double-click an issue to select that shape on the drawing.

Step 4: To ignore an issue, highlight the issue in the Issues window and select Ignore This Issue from the Ribbon.

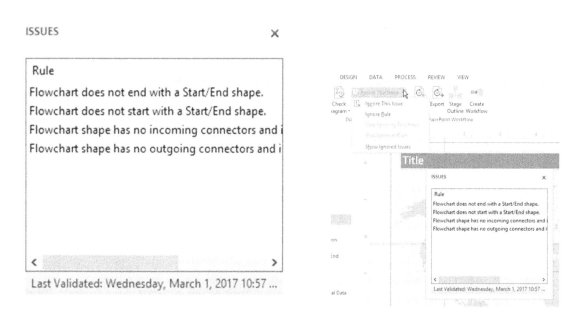

Using the Page Setup Group

The Page Setup Group includes options for determining your page orientation and size. You can have Visio auto size your drawing, based on the layout of the shapes it contains. The Page Setup dialog box allows you to customize your page setup.

Use the following procedure to change the page orientation.

Step 1: Select the Design tab from the Ribbon.

Step 2: Select Orientation.

Step 3: Select either Portrait or Landscape.

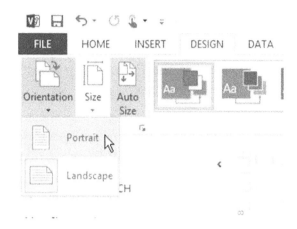

Use the following procedure to change the drawing size.

Step 1: Select the Design tab from the Ribbon.

Step 2: Select Size.

Step 3: Select a page size from the list.

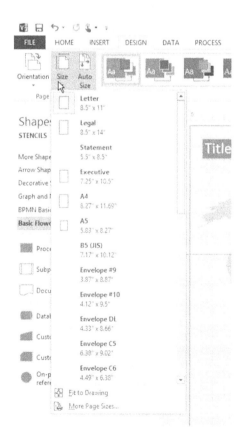

Page Setup dialog box.

Step 1: Select the Design tab from the Ribbon.

Step 2: Select the small square in the bottom right corner of the Page Setup group on the Ribbon to open the Page Setup dialog box.

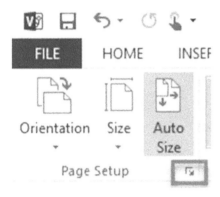

The Print Setup tab allows you to choose the printer paper, zoom level, and whether the gridlines should print.

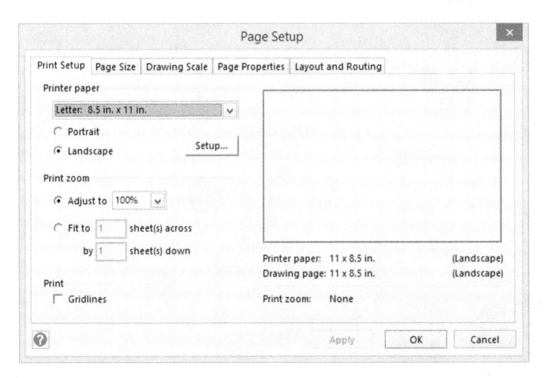

The Page Size tab allows you to choose the page size and orientation.

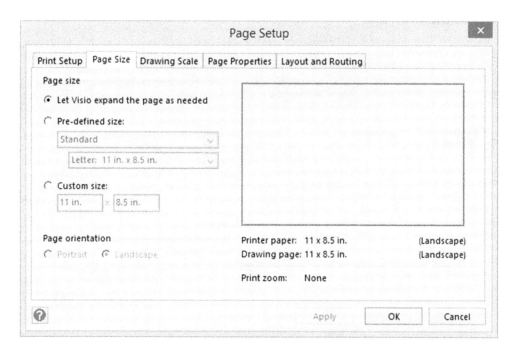

The Drawing Scale tab allows you to set the scale of the drawing.

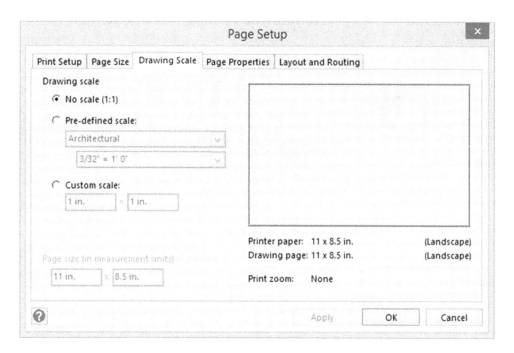

The Page Properties tab allows you to choose the name, type, background, and measurement units of the drawing.

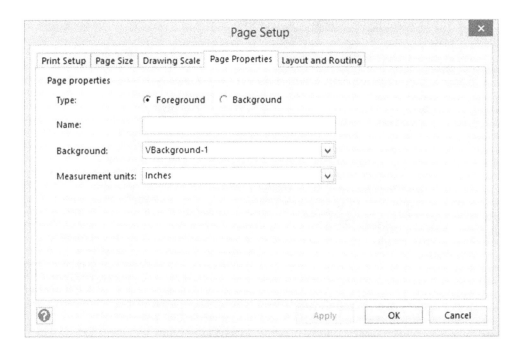

The Layout and Routing tab allows you to indicate routing and line jumps for your diagram.

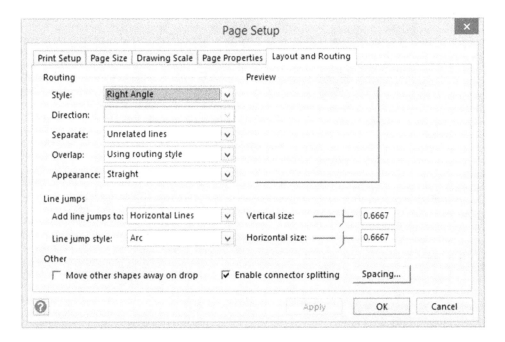

The Alt tab allows you to add a title and description for elements of the diagram. This allows for ADA accessibility.

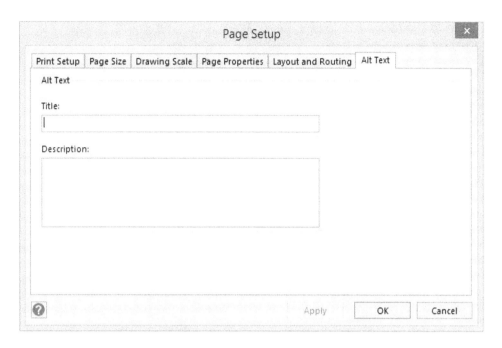

Saving Your Drawing as a Picture

You can save your drawing as any of the following formats to share with other viewers who may not have Visio installed on their computers:

- Web Drawing (for use on Visio Services with SharePoint)
- PNG
- EMF Metafile
- JPEG
- Scalable Vector Graphics
- XML Drawing
- HTML Page
- AutoCAD
- PDF
- XPS

Use the following procedure to save the drawing as a picture.

Step 1: Select the File tab on the Ribbon.

Step 2: Select the Save As item.

Step 3: Select Browse.

Step 3: Select the File Type under the Save As Type drop-down.

Step 4: Select the Graphic File Type.

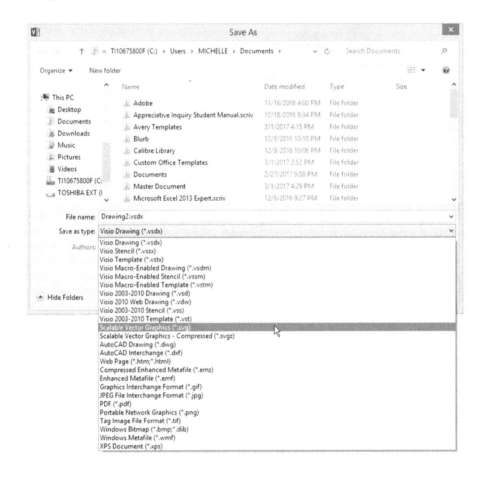

Printing Your Drawing

The Backstage View allows you to open a print preview, quick print using the default settings, or open the Print dialog box to set your print options.

Use the following procedure to open the Print dialog box.

Step 1: Select the File tab on the Ribbon.

Step 2: Select the Print item.

Step 3: Select Print.

E-mailing Your Drawing

Visio 2016 makes it easier than ever to share your files. You can attach the drawing to an E-mail in either the Visio format, PDF format, or XPS format. You can also email a link of the drawing if the drawing is saved to a shared folder.

Use the following procedure to email a drawing.

Step 1: Select the File tab on the Ribbon.

Step 2: Select the Share item.

Step 3: Select Send Using E-mail.

Step 4: Select Send as Attachment.

This chapter will provide the groundwork in some basic Visio terms before you start learning the procedures related to those terms. We'll also create some drawings based on Visio templates: a calendar, a map, a flowchart, and an organization chart. You'll also learn how to use perspective drawings.

Understanding Visio Definitions

A Drawing is the same as a File in Visio. The drawing contains all the elements that come together to help others visualize your idea. A drawing can contain multiple pages. It stores the stencils used on those pages.

Pages are just like pages in any other application. The point to remember is that drawings can contain multiple pages.

Layers work together to make up a page. We'll talk in more details about layers later in the course. For now, just remember that each page can have multiple layers, and different pages of a drawing can have different sets of layers.

Shapes are the objects in your drawing, whether one dimensional (lines) or two dimensional.

Stencils are containers for your shapes. Stencils don't do anything except store related shapes for easy access. You can have multiple stencils that you use in a drawing, including stencils you create.

Templates are the Visio starter drawings that include common stencils and shapes needed for that kind of drawing. They may also include layers to help you organize your drawing.

Creating Calendars

Use the following procedure to create a daily calendar in Visio:

Step 1; Select the File tab to open the Backstage View.

Step 2: Select New.

Step 3: Select the Schedule Template Category.

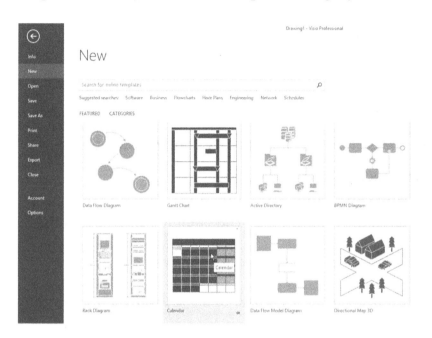

Step 4: Select Calendar.

Step 5: Select the measurement units.

Step 6: Select Create.

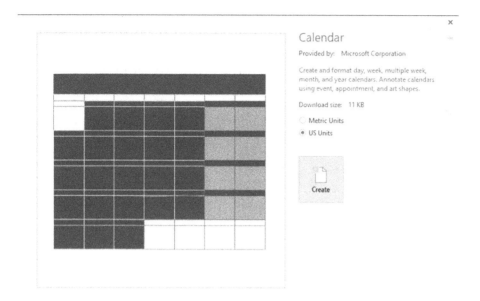

Calendar

Provided by: Microsoft Corporation

Create and format day, week, multiple week, month, and year calendars. Annotate calendars using event, appointment, and art shapes.

Download size: 11 KB

○ Metric Units
● US Units

Create

Look at the different shapes available in the Calendar template.

Drag a day, week, month, and appointment shape to the drawing one at a time and investigate the configuration options with each shape.

Step 1: Drag appointment to canvas.

Step 2: Right-click and select Configure to open the Configure window.

Step 3: Enter the Configure settings.

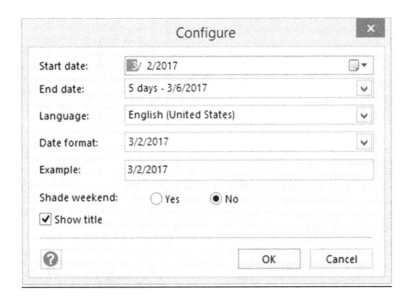

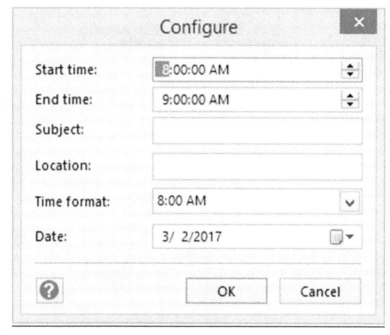

Creating Maps

Use the following procedure to create a directional map:

Step 1: Select the File tab to open the Backstage View.

Step 2: Select New.

Step 3: Select the Maps and Floor Plans Template Category.

Step 4: Select Directional map.

Step 5: Select the measurement units.

Step 6: Select Create.

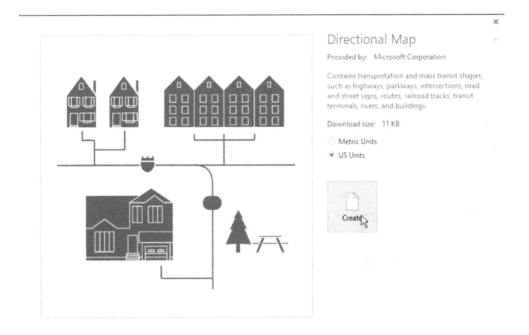

Look at the different shapes available in the map template, including the Metro, Recreation, and Transportation categories.

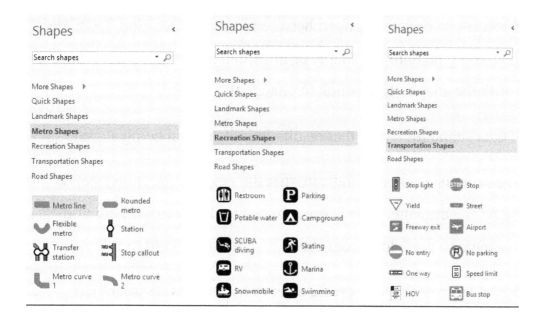

Drag shapes to the drawing to create your map.

Flowcharts Shapes

The following flowchart shapes are available:

Use the Terminator shape for the first and last step of your process.

Use the Process shape to represent a step in your process.

Use the Predefined process shape for a set of steps that combine to create a sub-process defined on another page of the drawing.

Use the Decision shape to indicate a point where the outcome of a decision dictates the next step. Although you can use multiple outcomes, often there are just two —yes and no.

Use the Document shape to represent a step that results in a document.

Use the Data shape to indicate information coming into the process from outside, or leaving the process. This shape can also be used to represent materials and is sometimes called an Input/Output shape.

The Flowchart shape can be interchanged between Process, Decision, Document, or Data. Any text you type onto the shape, or information you add to its Shape Data, remains with the shape. Right-click to change the shape.

Use the Stored data shape for a step that results in information being stored.

Use the On-page reference shape to indicate that the next (or previous) step is somewhere else on the drawing. This is great for large flowcharts where you would otherwise have to use a long connector, which can be hard to follow.

Use the Off-page reference shape to create a set of hyperlinks between two pages.

Use the Dynamic connector to draw a path around other shapes.

Use the Line-curve connector when you need adjustable curvature in your connector between shapes.

Use the Auto-height box for a text box that adjusts to accommodate your text. You can set the width by dragging the sides.

Use the Annotation shape for a bracketed text box to add comments about your flowchart shapes.

Use the Manual input shape to show a step where a person provides information to the process.

Use the Manual operation shape to show a step that must be performed by a person.

Use the Internal storage shape to represent information stored on a computer.

Use the Direct data shape to represent information stored so that any single record can be accessed directly. This represents how a computer hard-drive stores data.

Use the Sequential data shape to represent information stored in sequence, such as data on a magnetic tape. When data is stored in sequence, it must be retrieved in sequence.

Use the Card and Paper tape shape to represent a physical card or paper tape. Early computer systems used a system of punch cards and paper tape to store and retrieve data and to store and run programs.

Use the Display shape to represent information that is displayed to a person, usually on a computer screen.

Use the Preparation shape to indicate where variables are initialized in preparation for a procedure.

Use the Parallel mode shape to show where two different processes can operate simultaneously.

Use the Loop limit shape to mark the maximum number of times a loop can run before it must go on to the next step.

Use the Control transfer shape to indicate a step that goes to a step other than the typical next step when certain conditions are met.

Use the following procedure to create a flowchart:

Step 1: Select the File tab to open the Backstage View.

Step 2: Select New.

Step 3: Select the Flowchart Template Category.

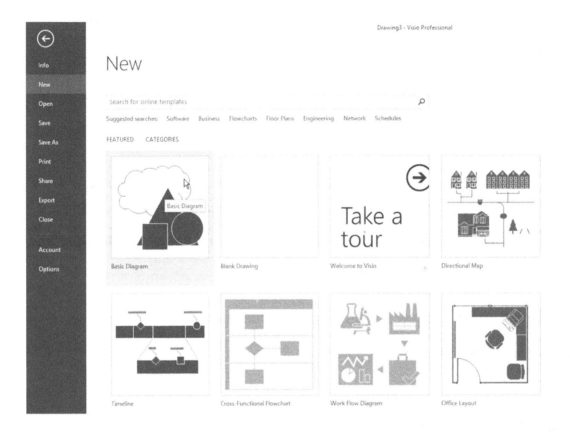

Step 4: Select Basic Flowchart.

Step 5: Select the measurement units.

Step 6: Select Create.

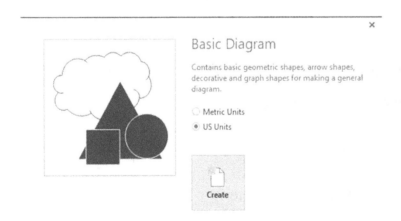

Look at the different shapes available in the flowchart template.

Drag shapes to the drawing to create your flowchart. You can use the connector tool on the Tools area of the Home tab on the Ribbon. Drag from a connection point on the first shape to a connection point on the second shape.

Creating Organization Charts

Use the following procedure to create an organization chart:

Step 1: Select the File tab to open the Backstage View.

Step 2: Select New.

Step 3: Search for Organization Chart.

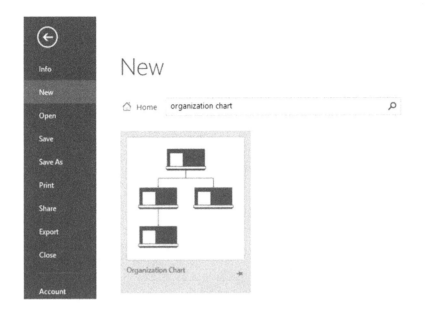

Step 4: Select Organization Chart.

Step 5: Select the measurement units.

Step 6: Select Create.

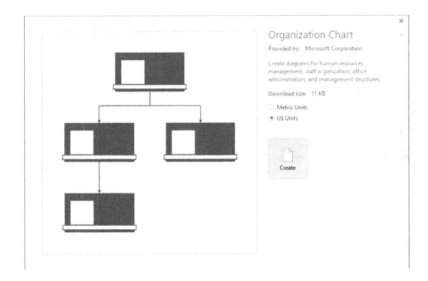

Look at the different shapes available in the flowchart template.

Drag shapes to the drawing to create your organization. Enter the information you want to include about each position or person in the box. To show a subordinate to that position, drag the next shape on top of it to automatically connect it.

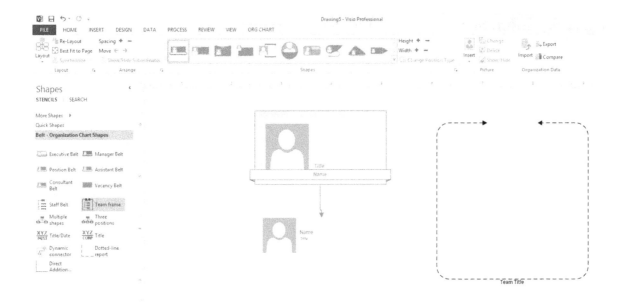

Using Perspective

Use the following procedure to create a block diagram with perspective:

Step 1: Select the File tab to open the Backstage View.

Step 2: Select New.

Step 3: Search for Block Diagram with Perspective.

Step 4: Select Block Diagram with Perspective.

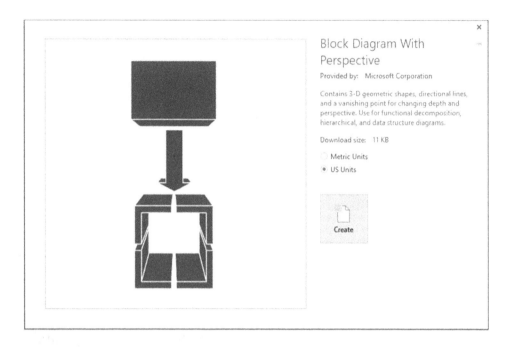

Step 5: Drag the Blocks with Perspective shapes onto the drawing.

Step 6: To change the perspective of a shape, drag the vanishing point (marked VP) to a new location.

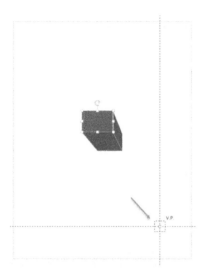

Step 7: To set the depth of a shape, right-click the shape you want to change and select Set Depth from the context menu.

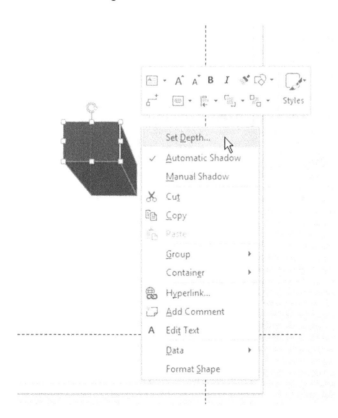

Step 8: Select the Depth from the drop-down list and select OK.

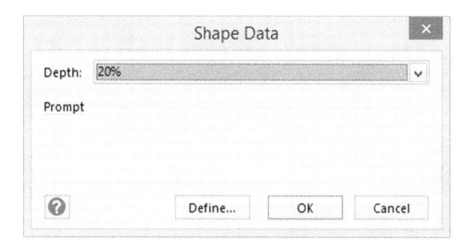

Network Diagrams

Use the following procedure to create a network diagram:

Step 1: Select the File tab to open the Backstage View.

Step 2: Select New.

Step 3: Search for Network Diagram.

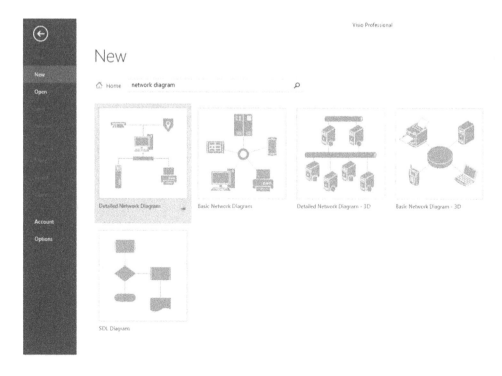

Step 4: Select Basic Network Diagram.

Step 5: Select the measurement units.

Step 6: Select Create.

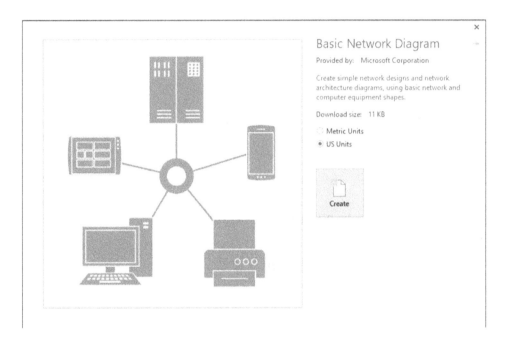

Look at the different shapes available in the network diagram template.

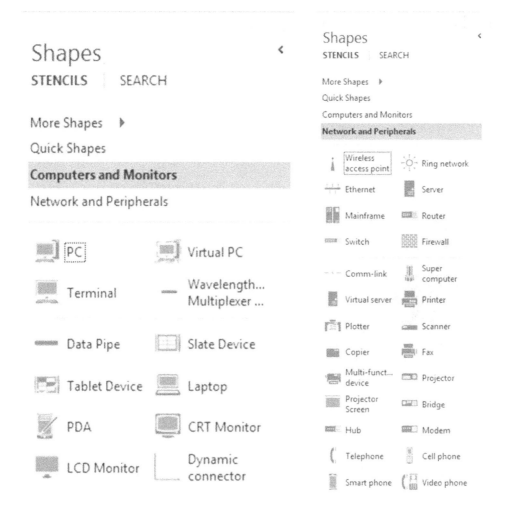

Drag shapes to the drawing to create your network diagram.

Marketing Diagrams

Use the following procedure to create a marketing diagram:

Step 1: Select the File tab to open the Backstage View.

Step 2: Select New.

Step 3: Search for Marketing Charts and Diagrams.

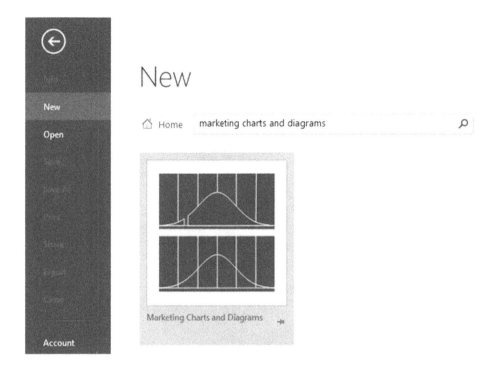

Step 4: Select Marketing Charts and Diagrams.

Step 5: Select the measurement units.

Step 6: Select Create.

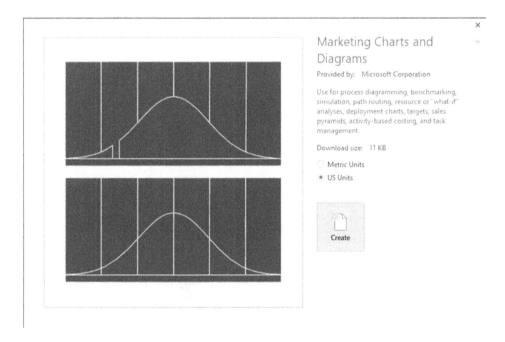

Look at the different shapes available in the marketing charts and diagram template.

Drag shapes to the drawing to create your marketing diagram.

Create Work Flow Diagrams

Use the following procedure to create a work flow diagram:

Step 1: Select the File tab to open the Backstage View.

Step 2: Select New.

Step 3: Search for Flow Chart.

Step 4: Select Work Flow Diagram.

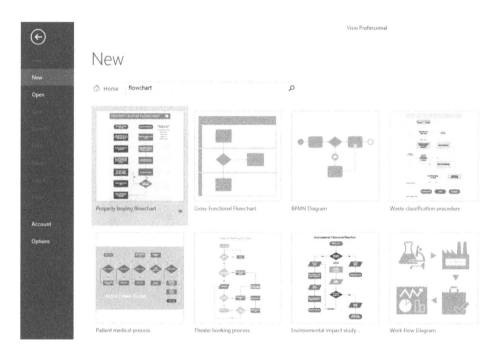

Step 5: Select the measurement units.

Step 6: Select Create.

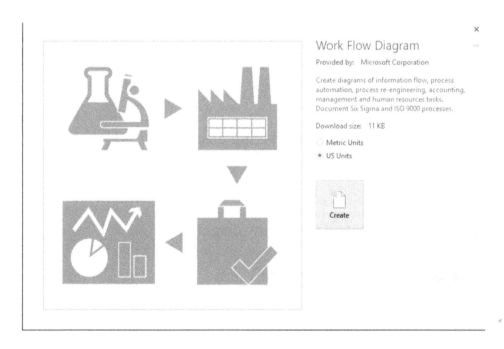

Look at the different shapes available in the Work Flow Diagram template. Use the scroll bar to see additional categories in the Shapes window.

Use the following procedure to create a cause and effect diagram:

Step 1: Select the File tab to open the Backstage View.

Step 2: Select New.

Step 3: Search for Cause and Effect Diagram.

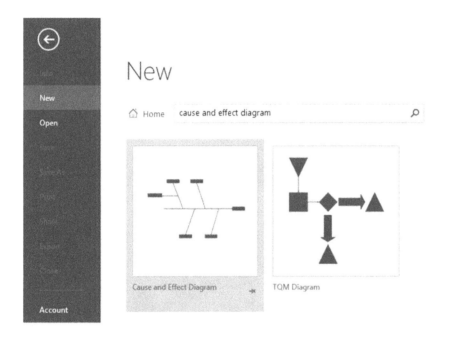

Step 4: Select Cause and Effect Diagram.

Step 5: Select the measurement units.

Step 6: Select Create.

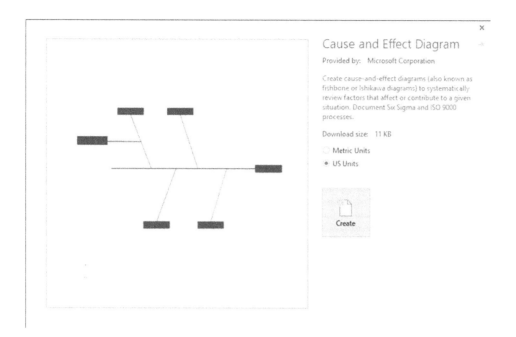

Look at the different shapes available in the Cause and Effect Diagram template.

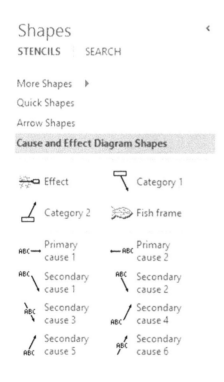

Step 7: On the drawing, select the spine and type text to describe the effect, problem, or objective.

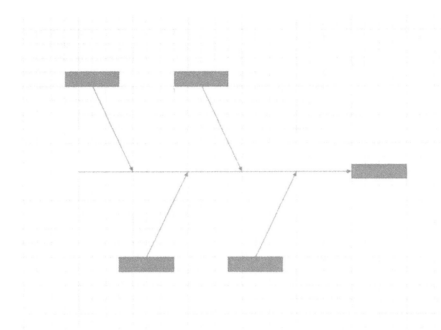

Step 8: Drag shapes to the drawing and position accordingly.

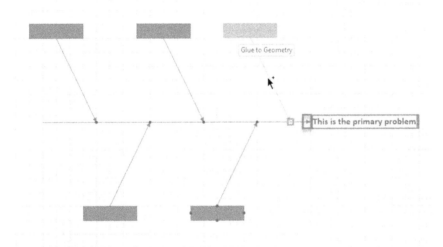

You can label the shapes, include primary and secondary cause shapes, and even rotate or flip shapes.

Project Management Diagrams

Use the following procedure to use the sample project management diagram:

Step 1: Select the File tab to open the Backstage View.

Step 2: Select New.

Step 3: Search for Pivot Diagram.

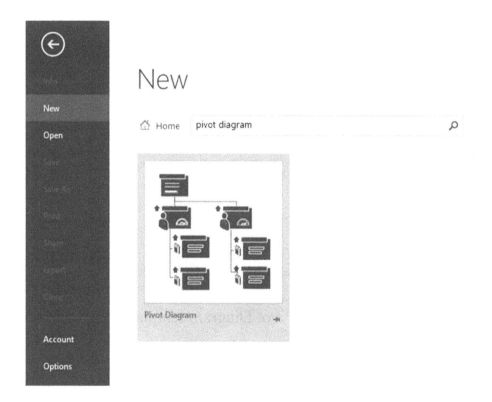

Step 4: Select Pivot Diagram.

Step 5: Select the measurement units.

Step 6: Select Create.

Step 7: Select the Data Source you want to use and then click Next.

Look at the different shapes available in the Pivot Diagram template.

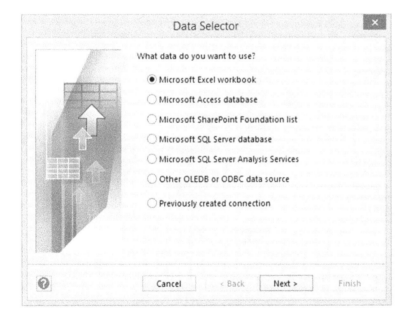

Gantt Charts

Use the following procedure to create a Gantt chart:

Step 1: Select the File tab to open the Backstage View.

Step 2: Select New.

Step 3: Search for Gantt Chart.

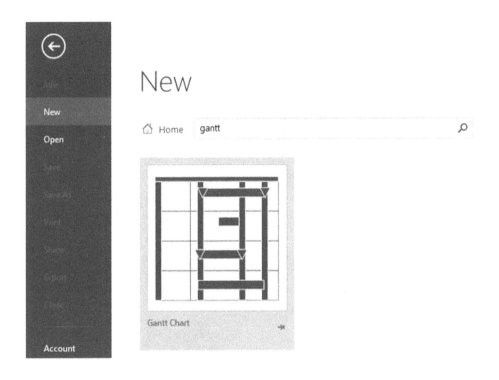

Step 4: Select Gantt Chart.

Step 5: Select the measurement units.

Step 6: Select Create.

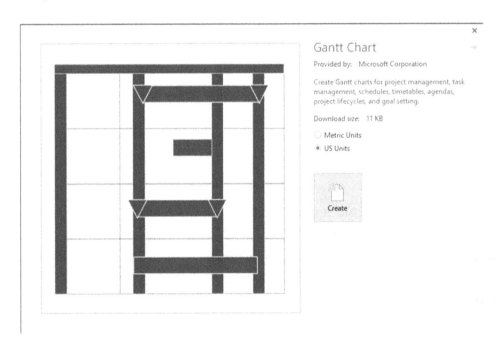

Visio displays the Gant Chart Options dialog box.

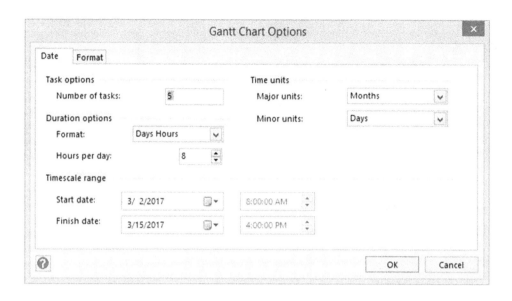

Step 7: Enter the Number of Tasks you want to represent on the chart. Enter the Time Units, Duration Options, and Timescale Range information.

Step 8: Review the options on the Format tab.

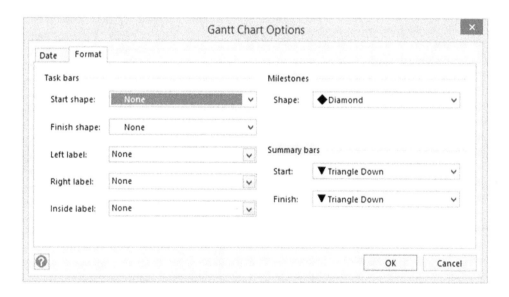

Step 9: Select OK.

Review your Gantt chart. Make changes in the fields as appropriate.

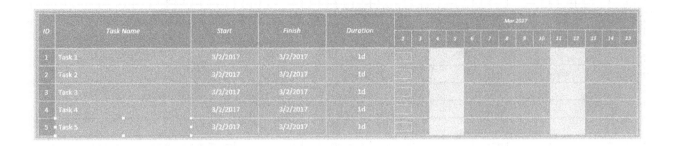

ID	Task Name	Start	Finish	Duration	Mar 2017													
					2	3	4	5	6	7	8	9	10	11	17	13	14	23
1	Task 1	3/2/2017	3/2/2017	1d														
2	Task 2	3/2/2017	3/2/2017	1d														
3	Task 3	3/2/2017	3/2/2017	1d														
4	Task 4	3/2/2017	3/2/2017	1d														
5	Task 5	3/2/2017	3/2/2017	1d														

Look at the different shapes available in the Gantt Chart template.

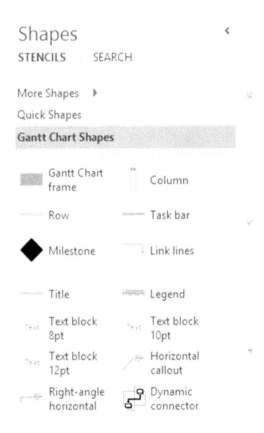

Review the Tools on the Gantt Chart tab of the Ribbon.

Hover the mouse over the tools to see screen tips.

Use the following procedure to create a PERT chart:

Step 1: Select the File tab to open the Backstage View.

Step 2: Select New.

Step 3: Search for PERT chart.

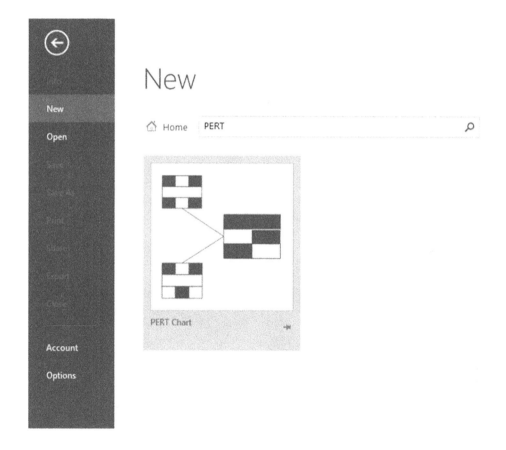

Step 4: Select PERT Chart.

Step 5: Select the measurement units.

Step 6: Select Create.

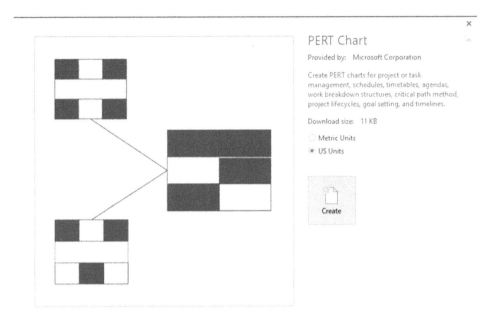

Look at the different shapes available in the PERT Chart template.

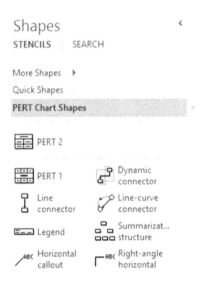

Drag shapes to the drawing to create your chart. The default text is only meant to guide you in your chart creation. You can select it to replace it with appropriate text.

Task Name	
Scheduled Start	Scheduled Finish
Actual Start	Actual Finish

This chapter will explain how to work with stencils. Remember that stencils are containers for shapes. First, we'll look at how to create and save a custom stencil. Then you'll learn how to add shapes to the stencil from other drawings or other stencils. Finally, we'll look at the controlling shape placement on your drawings using the Snap & Glue tools.

Creating Custom Stencils

Use the following procedure to create a custom stencil:

Step 1: Select the More Shapes option in the Shapes window. Select New Stencil.

Adding Shapes to the Stencil

Use the following procedure to add a shape to a stencil:

Step 1: Open the stencil where the shape you want to add will be stored. For this example, use the Favorites. Select More Shapes from the Shapes window. Select My Shapes. Select Favorites.

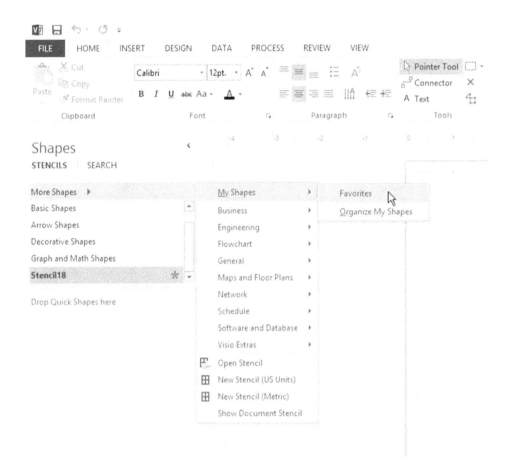

Step 2: Make sure that the stencil is editable by right-clicking the stencil title bar and select Edit Stencil from the context menu.

Step 3: Select the shape that you want to add to the stencil from the drawing page.

Step 4: To copy the shape, press the CTRL key while you drag the shape to the stencil.

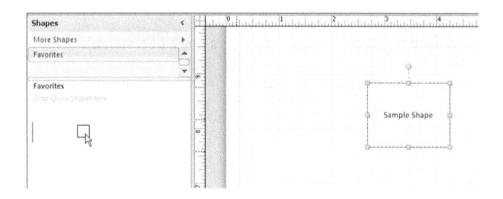

Step 5: The icon on the stencil is labeled as Master X, where X is a number corresponding to the number of shapes you have copied.

Use the following procedure to copy a shape from one stencil to another:

Step 1: Open the stencil that contains the shape you want to copy.

Step 2: Right click the shape that you want to copy and select Add to My Shapes. Select the stencil where you want to copy the shape.

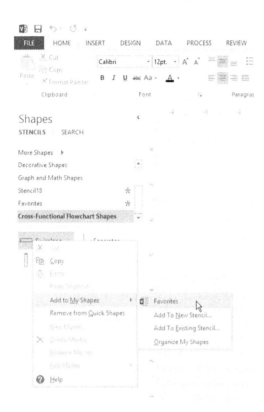

Saving the Stencil

Use the following procedure to save a stencil and save a copy of a stencil:

Step 1: Select the Save icon on the stencil title bar.

Step 2: To save a copy of the stencil, right-click the stencil title bar and select Save As from the context menu. Enter a name for the new stencil and select Save.

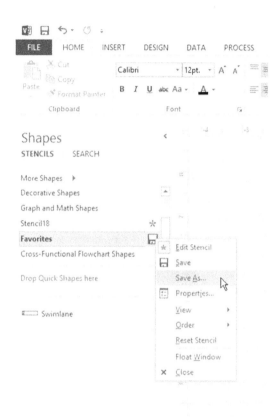

Controlling Shape Placement

Use the following procedure to snap shapes into position:

Step 1: Select the View tab from the Ribbon.

Step 2: Select the small square next to the Visual Aids group to open the Snap & Glue dialog box.

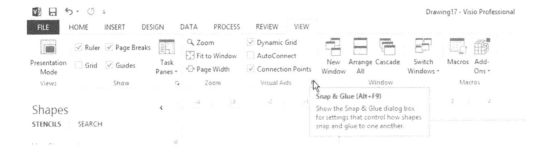

Step 3: Under Snap To, check the boxes to indicate which items you want to use when snapping shapes. These settings apply to all shapes in the drawing.

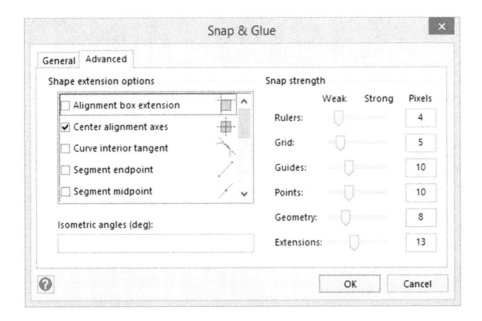

Step 4: Select OK.

Chapter 13 – Advanced Custom Shape Design

This chapter will explain the tools you need to customize shapes. We'll start with learning about the Quick Shapes area of the Shapes window. You'll learn how to create your own shapes and revise existing shapes. We'll also look at how to lock and protect shapes so that they won't get changed accidentally.

Using Quick Shapes

The Quick Shapes area is at the top of a stencil. Notice the thin divider between the Quick Shapes and the other shapes in that stencil.

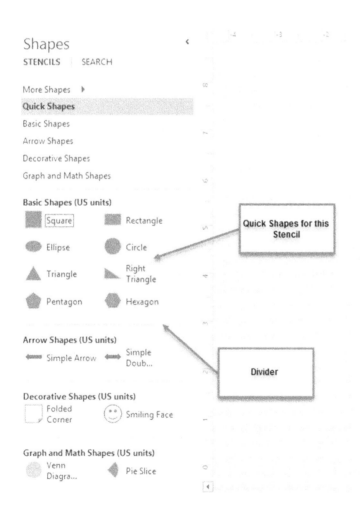

The following diagram shows the Quick Shapes stencil. In this example, there are three stencils open.

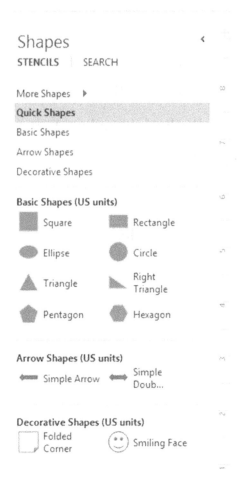

Note that the Quick Shape area allows you to use a few shapes from each stencil you have open. In this way, you won't have to switch between stencils as you work with your most used shapes.

Use the following procedure to move a frequently used shape to the Quick Shapes area of the stencil.

Step 1: Drag the Shape from the bottom part of the Stencil window to the top of that stencil. A small icon for the shape, plus a thin vertical bar shows where the shape will be placed in the Quick Shapes area

Note that you can use this same method to rearrange the order of the shapes in the stencil.

Now take another look at the Quick Shapes stencil. The shape you added to Quick Shapes in the Basic Shapes stencil is now shown in the Quick Shapes stencil.

Creating New Shapes

To create a shape on your Favorites stencil. Use the following procedure.

Step 1: To open the stencil where you want to store the new shape, select More Shapes from the Shapes window. Select My Shapes. Select Favorites or the name of your custom stencil.

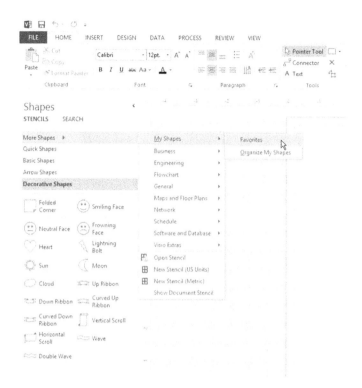

Step 3: Right-click the stencil title bar and select EDIT STENCIL from the context menu. The icon in the title bar changes to show that the stencil is editable.

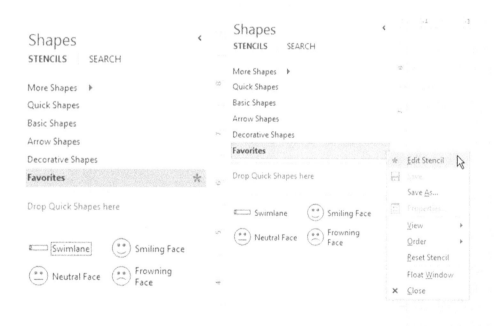

Step 2: Right-click the stencil window and select New Master from the context menu.

Step 4: Enter a Name for shape in the New Master dialog box.

Step 5: Enter a Prompt if desired.

Step 6: Select an Icon size to show on the stencil for the custom shape.

Step 7: Select Left, Center, or Right as the alignment.

Step 8: Check the Match master by name on drop box if desired.

Step 9: Check the Generate icon automatically from shape data box if desired.

Step 10: Select OK.

Step 11: Right-click the new blank shape icon in the stencil and select Edit Master from the context menu. Select Edit Master Shape.

Step 12: Visio opens a blank canvas for you to design the shape. You can draw the shape just as you can on a regular drawing page, using different stencil shapes, drawing using the drawing tools or pasting objects from another application.

Step 13: When you have finished drawing the shape, close the custom shape drawing window. Visio displays a dialog box asking if you want to save the changes to the shape. Select YES.

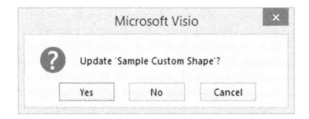

Step 13: Remember to Sᴀᴠᴇ your stencil.

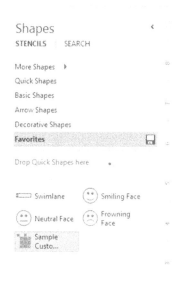

Revising Existing Shapes

Use the following procedure to edit a shape on a stencil:

Step 1: Open the stencil where the shape you want to edit is stored.

Step 2: If the stencil you opened is not editable, right-click the stencil title bar and select Edit Stencil from the context menu. The icon in the title bar changes to show that the stencil is editable.

Step 3: Right-click the stencil window and select Edit Master Shape from the context menu.

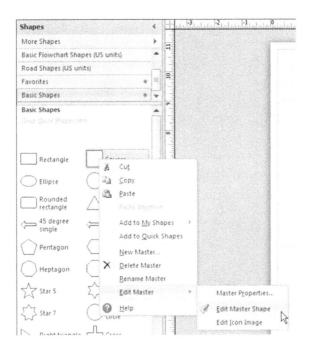

Step 4: Visio opens a separate canvas for you to edit the shape. You can use different stencil shapes, the drawing tools or pasted objects from another application.

Step 5: When you have finished drawing the shape, close the custom shape drawing window. Visio displays a dialog box asking if you want to save the changes to the shape. Select YES.

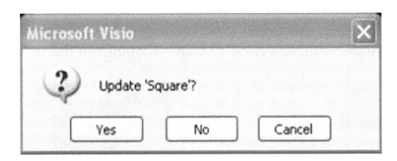

Step 6: Remember to Save your stencil.

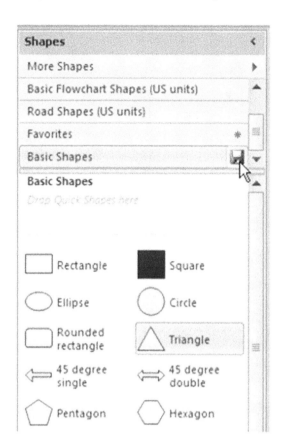

Locking and Protecting Shapes

Use the following procedure to display the Developer tab:

Step 1: Select the File tab to open the Backstage View.

Step 2: Select Options.

Step 3: Select the Advanced category.

Step 4: Scroll down to the General section.

Step 5: Check the Run in Developer Mode box.

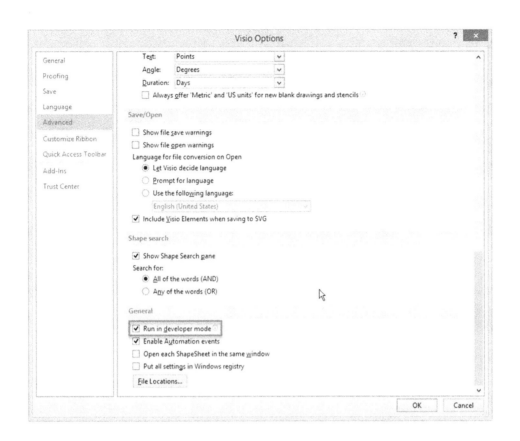

Step 6: Select Ok.

Step 7: Select the Developer tab.

Use the following procedure to protect a shape:

Step 1: Select the shape you want to protect.

Step 2: Select Protection from the Shape Design group on the Developer tab on the Ribbon.

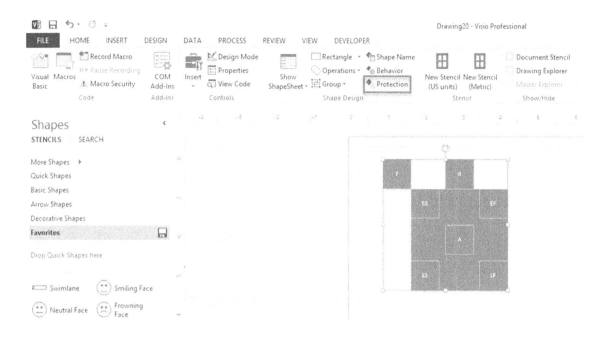

Step 3: Check the boxes to protect one or more of the following attributes of the selected shape.

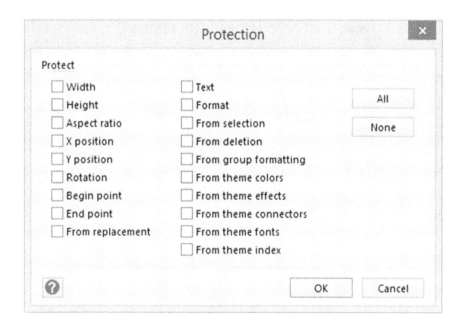

Step 4: Select OK.

Use the following procedure to lock a shape:

Step 1: Select the shape you want to protect.

Step 2: Select Protection from the Shape Design group on the Developer tab on the Ribbon.

Step 3: Check the From selection box.

Step 4: Select OK.

Step 5: Visio displays the Protection dialog box. Select OK to continue.

Step 6: Select Drawing Explorer from the Show/Hide group on the Developer tab of the Ribbon.

Step 7: Right-click the name of the drawing and select Protect Document from the context menu.

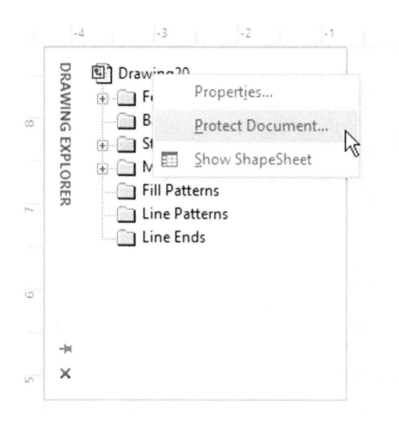

Step 8: Check the Shapes box to protect shapes from selection. Select OK.

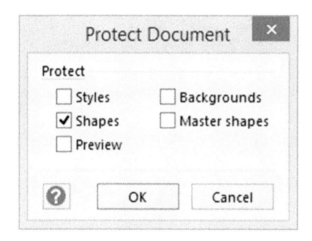

This chapter goes into how to use layers in Visio. You'll get an overview of layers, as well as learn how to create, remove, and rename layers. You'll also learn how to set layer properties and control shape placement. What if shapes were assigned to the wrong layer? You'll learn how to assign shapes to a new layer. You'll also learn how to assign color to a layer and will explain how to protect a layer from changes once you have it just like you want it. Finally, we'll look at how to print just the layers you want – one or more or all layers.

Understanding Layers

Layers are like have a drawing with multiple transparencies sitting on top of it. Each transparency shows a different aspect of the drawing. Using this concept, you can view or work with different layers at different times. Or you can view all the layers together.

Imagine that you are drawing an office layout. You could place the walls, doors, and windows on one layer. The electrical outlets could be on a separate layer. Then put the furniture on a third layer. You could lock the other two layers while you work with the shapes in the electrical system. That will keep you from accidentally moving the walls or furniture.

The Layer Properties Dialog Box

The Layer Properties dialog box.

Step 1: Select Layer from the Editing group on the Home tab of the Ribbon.

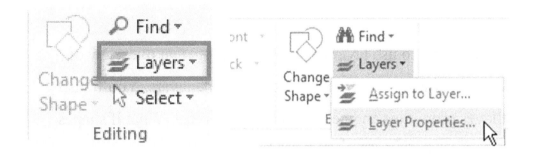

Step 2: Select Layer Properties from the Layers drop-down list to open the Layer Properties dialog box.

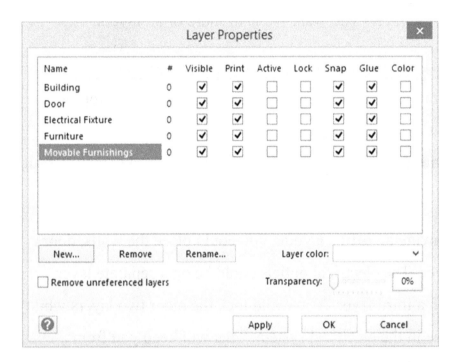

Working with Existing Layers

Use the following procedure to select shapes by layer:

Step 1: Choose the Select tool from the Editing menu on the Home tab of the Ribbon. Choose Select by Type.

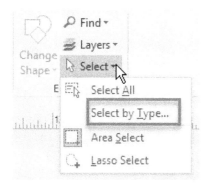

Step 2: In the Select by Type dialog box, select the Layer option.

Step 3: Check one or more boxes to indicate which layers' shapes you want to select. Only the layers that are currently displayed are available.

Step 4: Select OK.

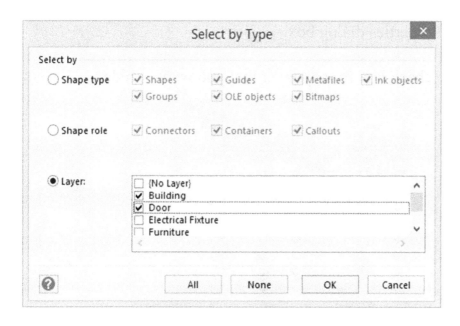

Visio shows the selected shapes on the currently displayed layers.

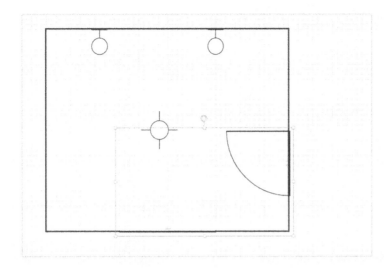

Hiding or Showing a Layer

Use the following procedure to hide a layer:

Step 1: Open the Layer Properties dialog box.

Step 2: Clear the Visible check box for any layer that you want to hide.

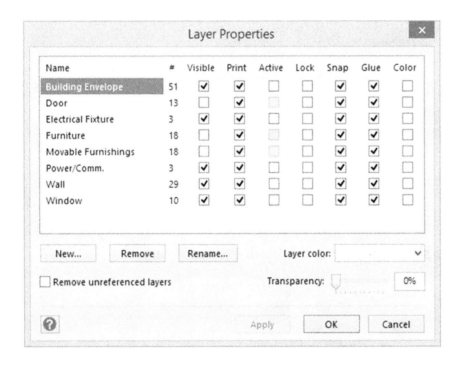

Step 3: Select OK.

Step 4: When you want to show the layers again, return to the Layer Properties dialog box and select the Visible checkbox next to the layers that you want to show.

Activating a Layer

Use the following procedure to activate a layer:

Step 1: Open the Layer Properties dialog box.

Step 2: Check the Active box next to any layers that you want to make active.

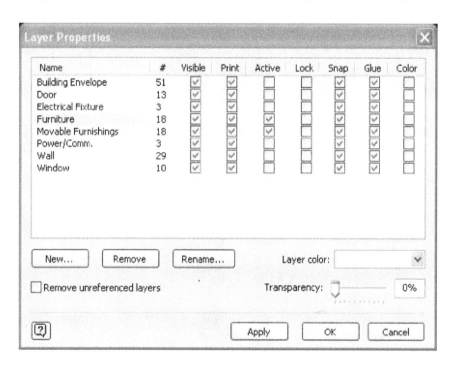

Creating Layers

Use the following procedure to add a layer:

Step 1: Open the Layer Properties dialog box.

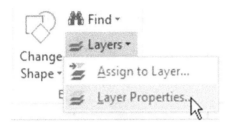

Step 2: Select New.

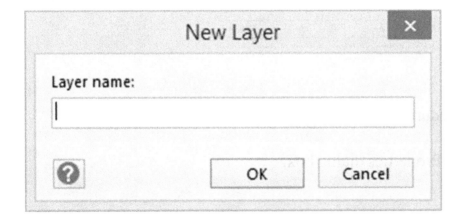

Step 3: Enter a name for your layer and select OK.

Step 4: Select OK in the Layer Properties dialog box.

Use the following procedure to rename a layer:

Step 1: Open the Layer Properties dialog box.

Step 2: Select the layer you want to rename.

Step 3: Select Rename.

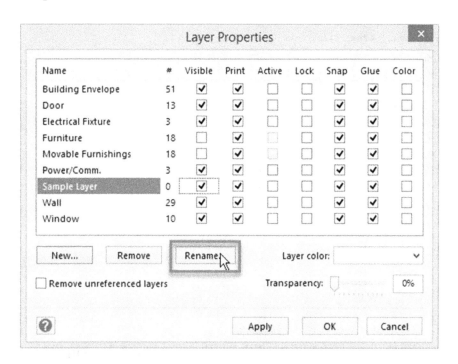

Step 4: In the Rename Layer dialog box, enter the new name and select OK.

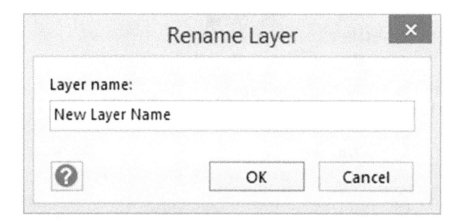

Step 4: In the Layer Properties dialog box select OK.

Use the following procedure to delete a layer:

Step 1: Open the Layer Properties dialog box.

Step 2: Select the layer you want to remove.

Step 3: Select Remove.

Step 4: If there are shapes on the selected layer, Visio displays the following message.

Step 5: Select Yes to continue.

Use the following procedure to assign a shape to a layer:

Step 1: Select the shape on your drawing. You can assign more than one shape at a time by holding the CTRL key down while selecting the shapes.

Step 2: Select Layer from the Home tab on the Ribbon.

Step 3: Select Assign to Layer.

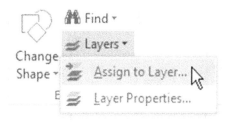

Step 4: On the Layer dialog box, check the boxes to indicate which layer(s) the shape should be assigned to. You can also select New to create a new layer and assign the shape at the same time.

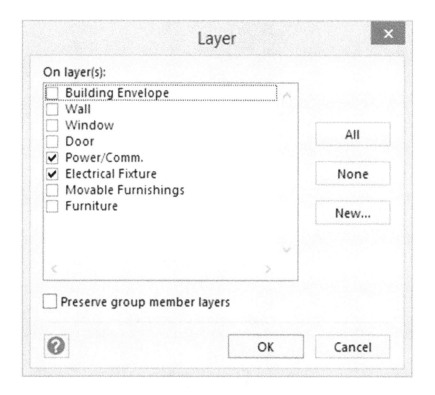

Use the following procedure to assign color to a layer:

Step 1: Open the Layer Properties dialog box.

Step 2: Select the layer you want to color.

Step 3: Select the Layer Color from the drop-down list.

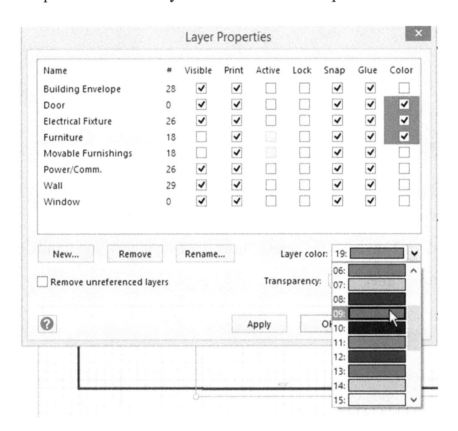

Step 4: To create a custom color, select More Colors from the Layer color drop-down list.

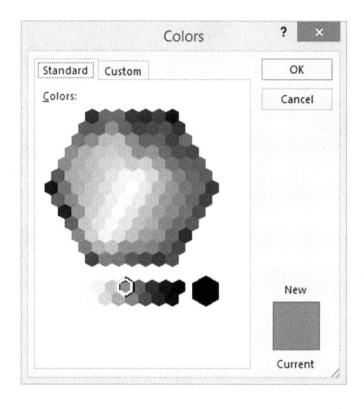

Step 5: Use the Colors dialog box to create your custom color and select OK.

Step 6: Select Apply.

Step 7: Select OK to close the Layer Properties dialog box.

Protecting a Layer from Changes

Use the following procedure to lock a layer:

Step 1: Open the LAYER PROPERTIES dialog box.

Step 2: Check the LOCK box next to any layers that you want to lock.

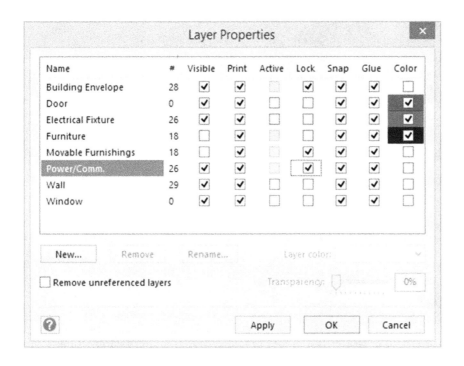

Printing Only the Layers You Want

Use the following procedure to print a layer:

Step 1: Open the Layer Properties dialog box.

Step 2: Check the Print box next to any layers that you want to print.

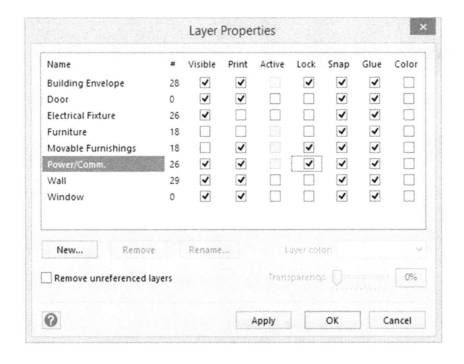

166

In this chapter, you'll learn how to work with Visio pages. First, we'll look at how to add pages to a drawing. Then, you'll learn how to arrange pages. You'll also learn how to work with background pages. Finally, you'll learn how to use hyperlinks to navigate between multiple pages in your drawing.

Adding Pages to a Drawing

Use the following procedure to add a new page to a drawing:

Step 1: Select the Insert tab from the Ribbon.

Step 2: Select the Blank Page tool.

Step 3: Select Blank Page.

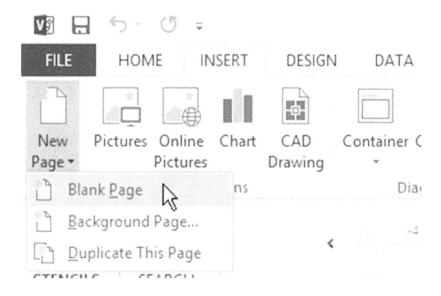

Visio adds a blank page with the same properties as the current page when you added the page.

You can also right-click the page tab at the bottom of the drawing window, and select Insert from the context menu.

Visio opens the Page Setup dialog box, open to the Page Properties tab.

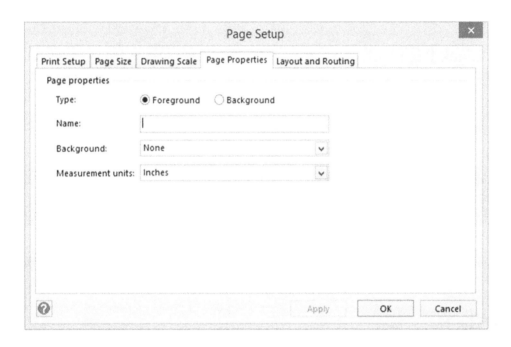

Step 1: Enter a Name for the page.

Step 2: If desired, select the Background page to apply to this page from the drop-down list.

Step 3: Select the Measurement units from the drop-down list.

Step 4: If you want to change other features of the page, use the Print Setup, Page Size, Drawing Scale, Layout and Routing, and Shadows tabs.

Step 5: Select OK.

Use the following procedure to reorder pages:

Step 1: Drag the page tabs into the new order.

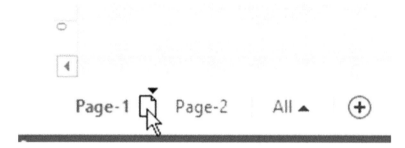

Step 2: You can also right-click the page tab and select Reorder Pages from the context menu.

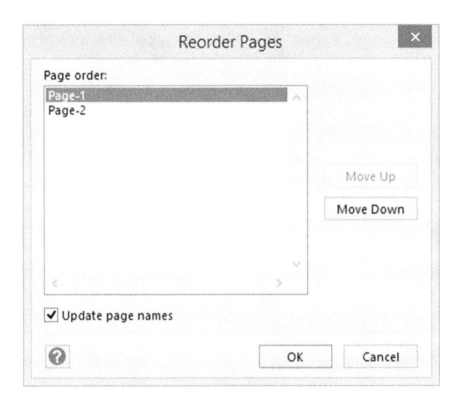

Step 3: Select Move Up or Move Down to rearrange the pages.

Step 4: Check the Update page names box if you want Visio to change the page names that include numbers.

Step 5: Select OK.

Working with Background Pages

Use the following procedure to add a background page:

Step 1: Select the Insert tab from the Ribbon.

Step 2: Select the Blank Page tool.

Step 3: Select Background Page. Or you can right-click a Page tab and select Insert from the context menu.

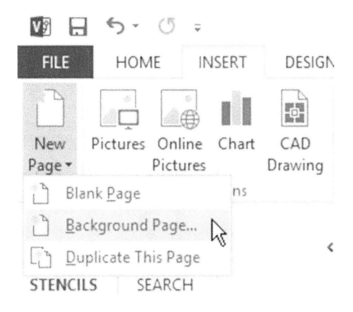

Step 4: In the Page Setup dialog box, select the Background option.

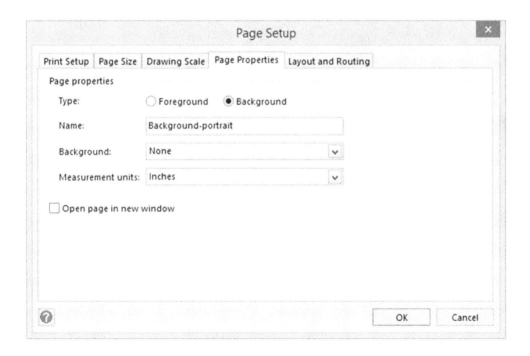

Step 5: Select OK.

Use the following procedure to assign a new background page to a page in a drawing:

Step 1: Right-click the Page tab for the page that you want to reassign.

Step 2: Select Page Setup from the context menu.

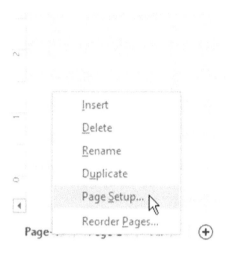

Step 3: On the Page Properties tab, select the new Background page from the drop-down list.

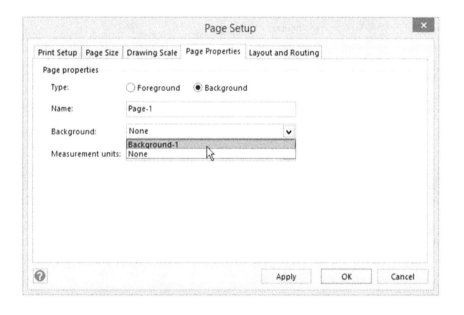

Step 4: Select Apply or OK.

Add shapes to the background page and then review the results on the pages that use that background page.

Hyperlinking Between Pages

Use the following procedure to add a hyperlink to another page in the drawing:

Step 1: Select the page or shape where you want to add the hyperlink.

Step 2: Select the Insert tab on the Ribbon.

Step 3: Select Hyperlink.

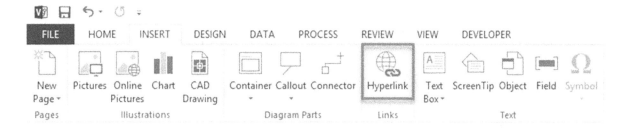

Step 4: Select Browse next to Address to locate the file you want to use, even though it is the same file.

Step 5: Select Local file

Step 6: Select the location of the current file and select Open.

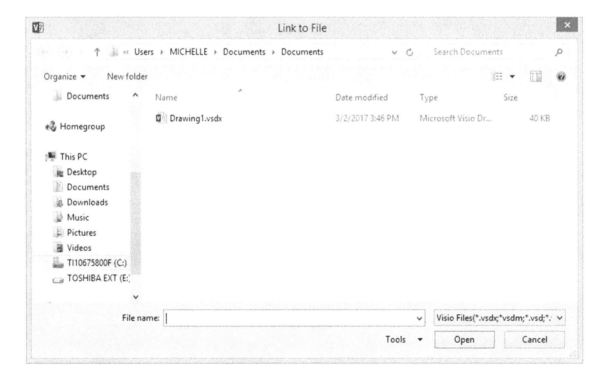

Step 7: Select Browse next to Sub-address to select the page.

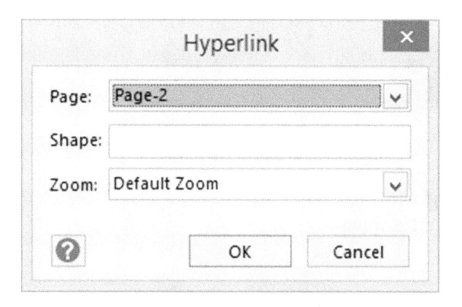

Step 8: Select the Page from the drop-down list.

Step 9: Enter the Shape, if appropriate.

Step 10: Select the Zoom level from the drop-down list.

Step 11: Select OK to close the dialog boxes.

When you hover the mouse over the link, the cursor shows that there is a link, as well as a screen tip for the link location. You can follow the link by pressing the Ctrl key while clicking the link or by right-clicking the shape and selecting the link from the context menu.

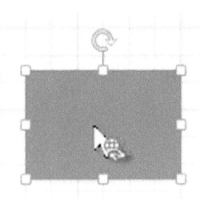

Drawing1.vsdx
Ctrl + Click to follow link

This chapter explains how to draw with precision using the Size and Position window. We'll look at viewing area measurements for your shapes. You'll also learn how to set the drawing scale on your drawings. Finally, we'll work with building plan layouts to practice your drawing skills.

Drawing with Precision

The Size and Position dialog box. Use the following procedure.

Step 1: Select the View tab from the Ribbon.

Step 2: Select Task Panes.

Step 3: Select Size & Position Window.

Step 4: The Size and Position window opens in the bottom left corner of the screen.

SIZE & POSITION...		
	X	5.25 in.
	Y	5.8636 in.
	Width	1.5 in.
	Height	1 in.
	Angle	0 deg.
	Pin Pos	Center-Center

SIZE & POSITION...		
	X	6.5 in.
	Y	2.75 in.
	Width	2.25 in.
	Height	1.5 in.
	Angle	0 deg.
	Pin Pos	Center-Center

The Size and Position window shows different information, based on the type of shape you have selected. These examples show an arrow, a square, and a line. If different types of shapes are all selected, or nothing is selected, the window shows No Selection.

In each of the available fields, you can enter a precise measurement to get the shape exactly like you want it. You can also enter formulas, such as +1.5 ft. or -90 deg. This is helpful if you want to apply the same change to multiple similar shapes.

The Pin Position field controls where the "pin" on the shape that controls the rotation is located. You can select a new option from the drop-down list.

Practice changing some of the Size and Position window measurements on their shapes to see the results.

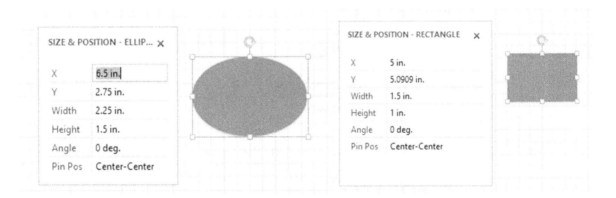

Use the following procedure to change the drawing scale:

Step 1: Display the page where you want a different drawing scale.

Step 2: Select the Design tab.

Step 3: Select the small square next to the Page Setup group.

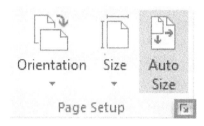

Step 4: Select the Drawing Scale tab.

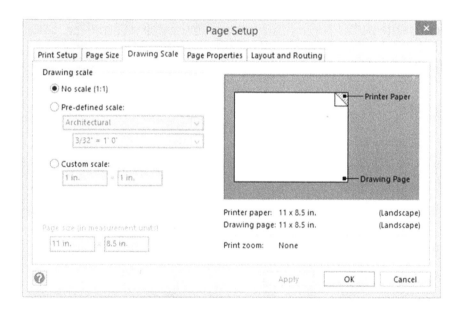

Step 5: Select Pre-defined scale and select a pre-defined type and scale from the drop-down lists. Or select Custom scale and enter the scale.

Step 6: To change the measurement units, select the Page Properties tab.

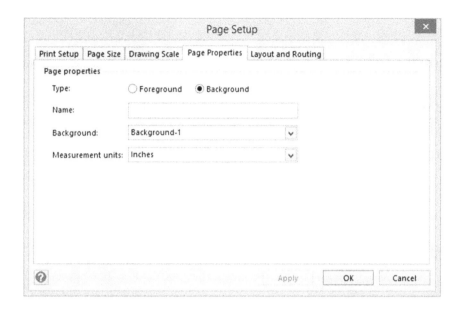

Step 7: Select the new Measurement units from the drop-down list.

Step 8: Select Apply to save your changes.

The drawing shows the new settings. Shapes might appear larger or smaller, but their real-world size does not change. Rulers show the new measurement units.

Working with Building Plan Layouts

Use the following procedure to create a floor plan:

Step 1: Select the File tab to open the Backstage view.

Step 2: Select New.

Step 3: Search for Floor Plans.

Step 4: Select Floor Plan.

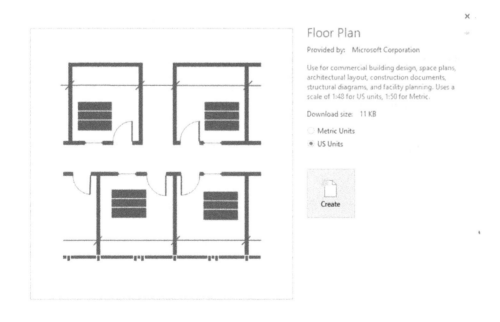

Step 5: Use the Walls, Shells, and Structure shapes to create the basic exterior wall structure, interior walls, and any other structure elements, doors, and windows. Then you can add electrical symbols and dimension lines.

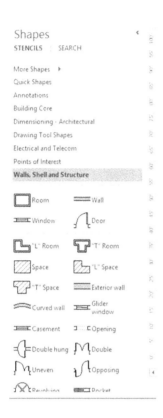

Use the following procedure to insert a CAD floor plan:

Step 1: Create a new floor plan (as in the previous procedure).

Step 2: Select the Insert tab from the Ribbon.

Step 3: Select CAD Drawing.

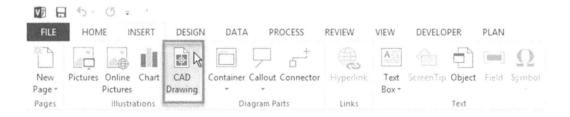

Step 4: Navigate to the AutoCAD Drawing you want to insert and select Open.

Step 5: To accept the size and location of the CAD drawing, select OK. You can resize it, change its scale, or move it once it is in the drawing. You can lock the layer that contains the CAD drawing so that it is not accidentally changed.

In this chapter, we'll look at using data in your drawings. First, we'll look at how you can use data with graphics to create a professional diagram that combines textual and graphical elements to illustrate your data. Then, you'll learn how to use the Data Selector Wizard to connect your drawing to an external data source. You'll learn how to apply a data graphic to your drawing. Finally, you'll learn how to edit the data graphic to get it looking just like you want.

About Data Graphics

Data graphics can enhance your shapes to show data the shapes contain. You can connect your drawings to data, either by entering the data in Visio, or by connecting your drawing to an external data source. Once your drawing is connected to a data source, you can add data graphics to your drawing either by adding new shapes based on the data, or by applying the data to existing shapes. A data graphic combines the following elements:

- Text
- Data bar
- Icons
- Color

Quick import of an Excel workbook

The most popular data source to import is an Excel workbook. That's why there is a Quick Import button for this source.

Step 1: Click Quick Import on the Data tab.

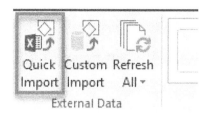

Step 2: Click Browse, and then select the workbook you want to import.

Step 3: Click Open, and then click Done.

Step 4: If the Import to Visio box and the Excel program appear, click the sheet tab where your data is, and then drag to select your data. Make sure to include any headers above the columns. Then, click Import in the Import to Visio box, and then click Done.

Custom import process for all data sources

If the Quick Import option did not work as you expect, or you have another data source to import, you can do a custom import of the data.

Step 1: Click Custom Import on the Data tab.

Step 2: Choose a data source on the first page of the Data Selector wizard and then click Next.

Step 3: Click Browse, and browse to the source you want to import, and then click Next.

Step 4: Select the check box for the column containing the unique values on the Configure Refresh Unique Identifier page. If you refresh the imported data

186

later, the unique identifier will enable Visio to find the updated row in the source, retrieve it, and then update the diagram. Click Next.

Step 5: The External Data window appears with the imported data shown in a table after you click Finish on the last page of the Data Selector wizard.

Applying Data Graphics

Use the following procedure to drag a data row onto the drawing:

Step 1: Drag a row from the External Data window onto the drawing. Visio automatically applies the shape you selected in the Shapes window.

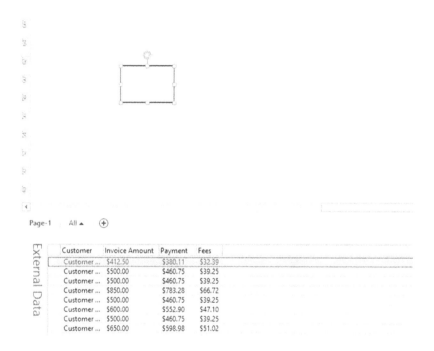

Step 2: Notice the chain icon in the External Data window, indicating that the row is linked to a shape.

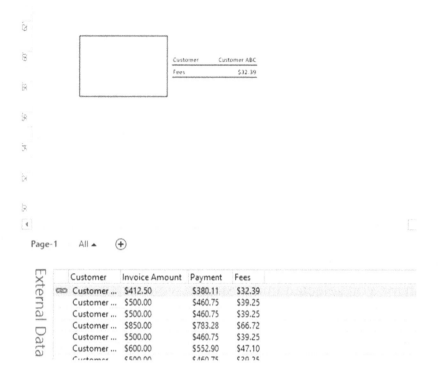

Customer	Invoice Amount	Payment	Fees
Customer ...	$412.50	$380.11	$32.39
Customer ...	$500.00	$460.75	$39.25
Customer ...	$500.00	$460.75	$39.25
Customer ...	$850.00	$783.28	$66.72
Customer ...	$500.00	$460.75	$39.25
Customer ...	$600.00	$552.90	$47.10
Customer	$500.00	$460.75	$39.25

Use the following procedure to open the Data Graphic task pane:

Step 1: Select Task Pane from the View tab on the Ribbon.

Use the following procedure to apply a different data graphic to a shape with data:

Step 1: Select the shape in the drawing.

Step 2: Select the Data tab from the Ribbon.

Step 3: Select Data Graphics.

Step 4: Select the data graphic from the Available Data Graphics gallery. You can see a live preview of the new data graphic before you apply it.

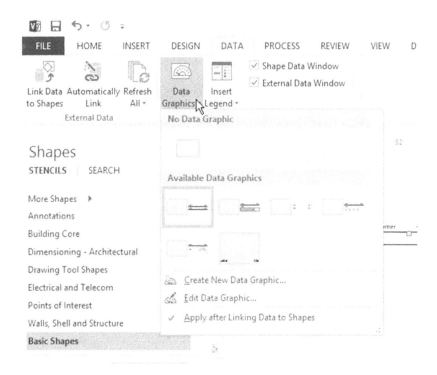

Editing Data Graphics

Use the following procedure to edit a Data Graphic:

Step 1: Right-click the shape and select Data from the context menu. Select Edit Data Graphic.

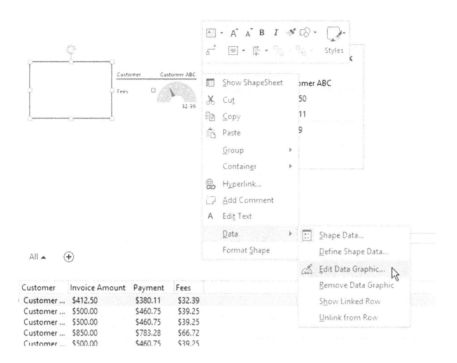

Step 2: In the Edit Data Graphic dialog box, you can choose a new data field to display for the items already added to this data graphic by choosing a new field from the drop-down lists. You can also change the display position or change the display options.

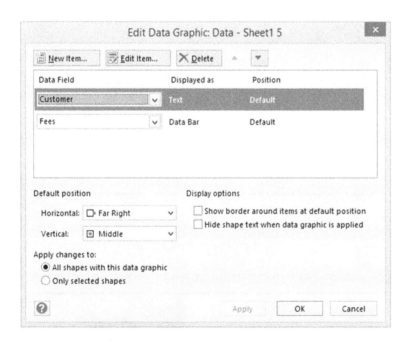

Here's what the data graphic looks like with a few changes.

It's still not what I'm after. So, let's edit the item to further refine it.

Step 3: Select the item that you want to modify and select Edit Item. Let's start with the Text item.

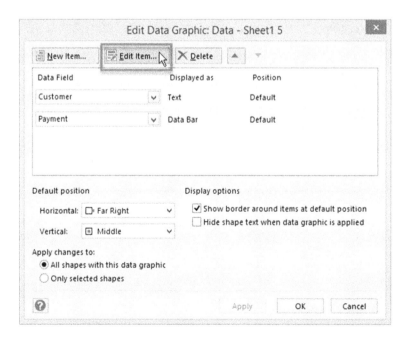

Step 4: The Edit Text dialog box opens to help determine how to format the text. You can select a new column from the Data field drop-down list, if desired. However, I want to leave it as Customer.

Step 5: Let's select a new Callout format from the drop-down list.

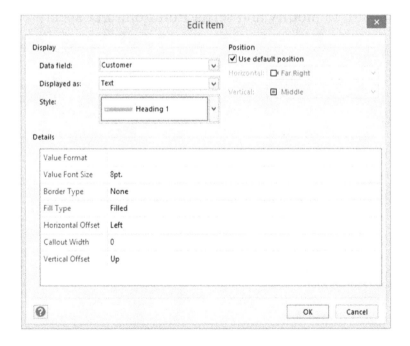

Step 6: You can customize the Callout position, or just leave the Use default position check box checked.

Step 7: In the Details area, you can customize the data graphic further. We can leave the defaults for now.

Step 8: Select OK to save your changes to the Text item.

Step 9: Now let's adjust the data bar part of the graphic. Select the Data Bar item and select Edit Item.

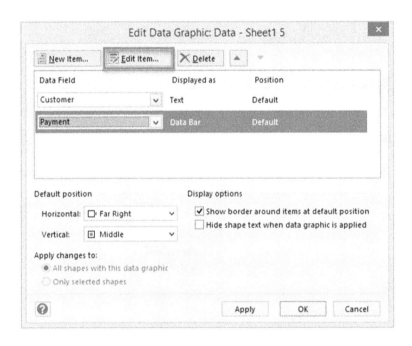

Step 10: The Edit Data Bar dialog box opens to help you format the data graphic. It works in a similar way to the Edit Text dialog box. Just select the Data field and Callout format from the drop-down lists. You can leave the defaults on everything else. I want to select the Speedometer data bar format. Select OK to continue.

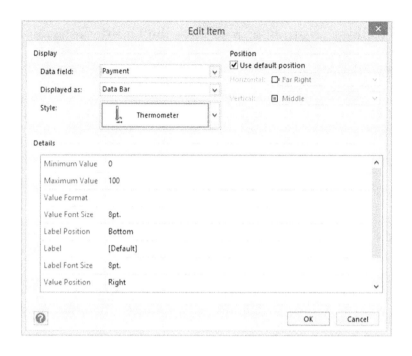

Step 11: Select OK to close the Edit Data Graphic dialog box.

Here is the updated data graphic.

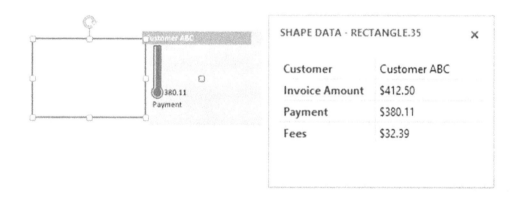

Use the following procedure to add a new item to the data graphic:

Step 1: Open the Edit Data Graphic dialog box as you did in the previous procedure.

Step 2: Select New Item.

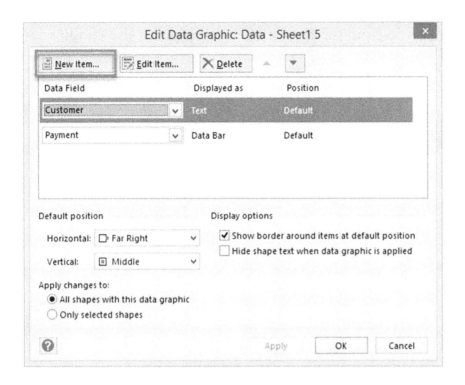

Step 3: The New Item dialog box opens. Choose a Data field from the drop-down list.

Step 4: Choose an element type from the Displayed as drop-down list. In this example, we'll choose Icon set.

Step 5: The New Icon Set dialog box opens to help you format the data graphic. It works in a similar way to the Edit Text and Edit Data Bar dialog boxes. Just select a Style from the drop-down list. You can leave the defaults on the position. You'll need to add rules for the icons that you want to use. Select an option from the drop-down list (or select [Not Used]) and enter a value, an expression or a date. Select OK when you have finished.

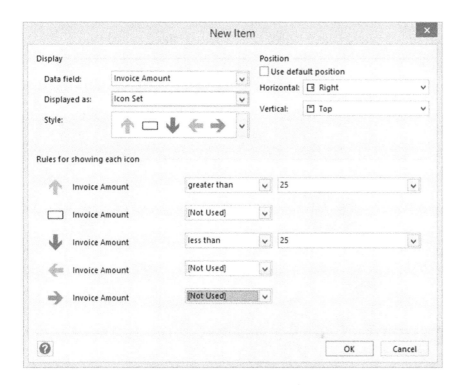

Step 6: Select Apply to preview the changes. Select OK to close the Edit Data Graphic dialog box.

Data Graphic Legends

Use the following procedure to insert a data graphic legend:

Step 1: Select the Data tab from the Ribbon.

Step 2: Select Insert Legend.

Step 3: Select the type of legend you want to use.

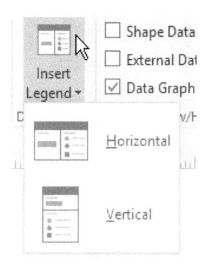

Shape Data

External Dat

☑ Data Graph

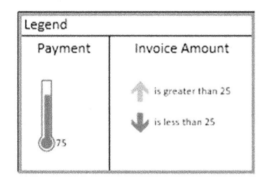

Horizontal

Vertical

Legend	
Payment	Invoice Amount
75	⬆ is greater than 25 ⬇ is less than 25

Chapter 18 – The ShapeSheet

In this chapter, we'll look at the ShapeSheet. You'll learn how to view the ShapeSheet and how to modify ShapeSheet data. You'll also learn how to use formulas in the ShapeSheet.

Viewing the ShapeSheet

Use the following procedure to view the ShapeSheet

First, you'll need to display the Developer tab, if it isn't already displayed.

Step 1: Select the FILE tab to open the Backstage View.

Step 2: Select OPTIONS.

Step 3: Select the ADVANCED category.

Step 4: Scroll down to the GENERAL section.

Step 5: Check the RUN IN DEVELOPER MODE box.

Step 6: Select OK.

If the Developer tab is already displayed, start here.

Step 7: If you want to display the ShapeSheet for the shape or the page, make sure you select the desired shape first or make the page you want to show active.

Step 8: Select the DEVELOPER tab.

Step 9: Select Show ShapeSheet.

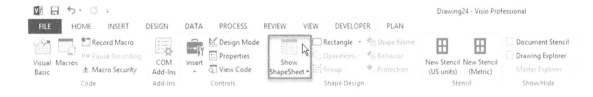

Step 10: Select Shape, Page, or Document.

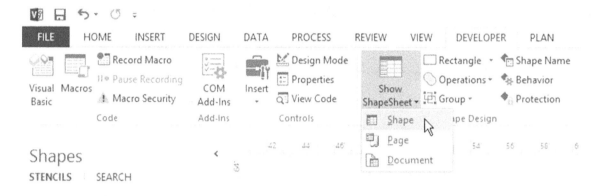

The following diagram shows the ShapeSheet Tools Design tab.

If you select a shape before opening the ShapeSheet, the ShapeSheet shows data for only the selected shape.

The Shape ShapeSheet data includes values for the shape, the user-defined cells (or data), the connection points, Geometry data, Protection data, Miscellaneous, the Line Format data, the Fill format data, the Character, Paragraph, Tab and Text block formatting, Events, Image Properties, Glue Info and Shape Layout.

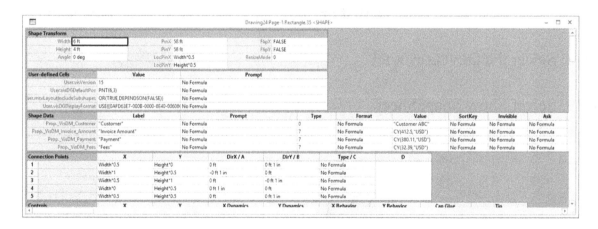

The Page ShapeSheet data includes values for the page properties, the Page layout, the Rule & grid properties and the Print properties.

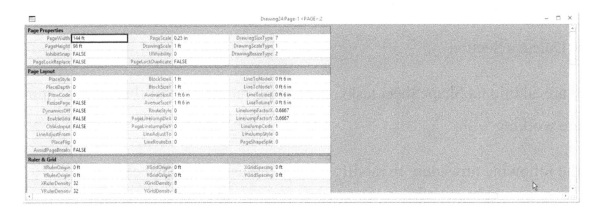

The Document ShapeSheet includes values for the document properties and user-defined cells.

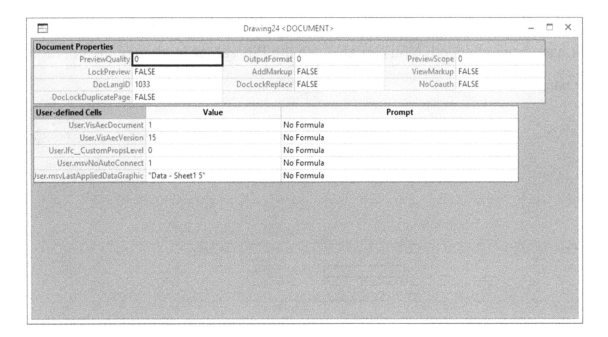

Modifying ShapeSheet Data

In addition to modifying values in the ShapeSheet data fields, you can also perform the following tasks:

- Add a New Row (for some sections)

- Add a Section
- Change a Row Type
- Delete a Row
- Delete a Section

There are additional ShapeSheet tasks, depending on the type of shape or section you are working with.

Use the following procedure to modify a shape by using the ShapeSheet:

Step 1: Select the shape you want to change.

Step 2: Display the ShapeSheet for the selected shape.

Step 3: Enter new values in the desired fields. In this example, the width and height have been changed. You can also use the Pin values to change the position of the shape.

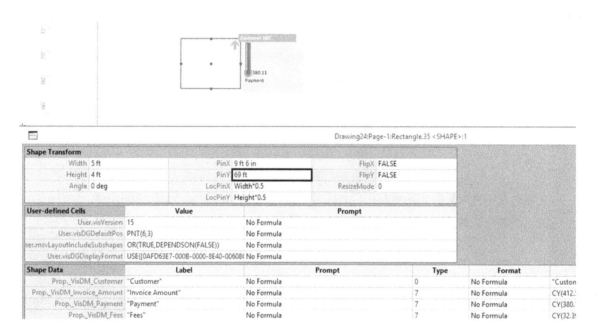

Use the following procedure the options for additional modifications to the ShapeSheet:

Step 1: Right-click in the ShapeSheet to see the options.

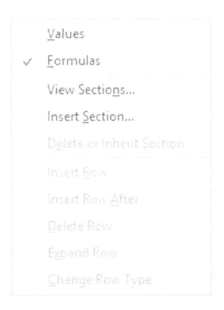

Investigate the options (View Sections, Insert Section, Delete Section, Insert Row, Insert Row After, Delete Row, Expand Row and Change Row Type), depending on the level of the students and the time available. These options are also available from the ShapeSheet Tools Design tab.

Using a Formula in the ShapeSheet

Use the following procedure to enter a formula into a ShapeSheet cell:

Step 1: Click in the ShapeSheet window to make it active.

Step 2: Make sure that formulas are displayed. To check, select Formulas from the ShapeSheet Tools Design tab on the Ribbon.

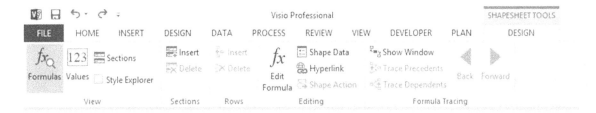

Step 3: Place your cursor in the cell where you want the formula.

Step 4: Use the Formula bar to enter your formula.

Step 5: Select Accept (or press the Enter key) to complete the formula. Notice the cell may contain the result of the formula instead of the formula itself.

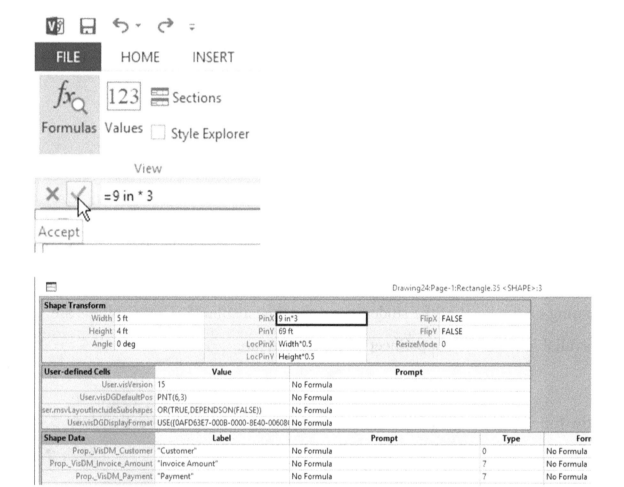

You can also use the Edit Formula tool from the ShapeSheet Tools Design tab.

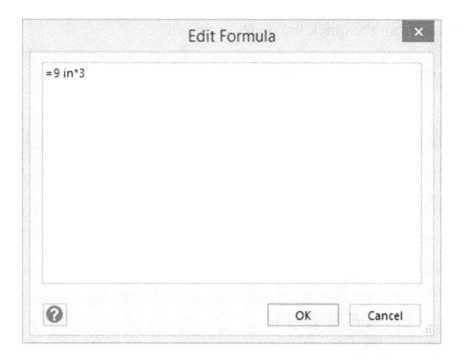

Use the following procedure to insert a function into a formula:

Step 1: Double-click the cell where you want to use a formula with a function.

Step 2: Place the insertion point where you want to paste the function.

Step 3: Type = to begin your formula.

Step 4: Type the first few letters of the function you want to use.

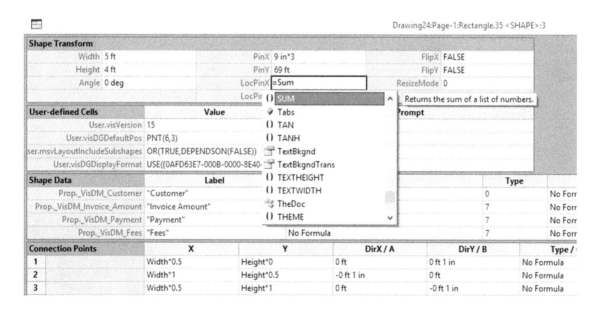

Step 5: Select the Function from the quick list under your cell.

Step 6: Visio helps you with the function arguments.

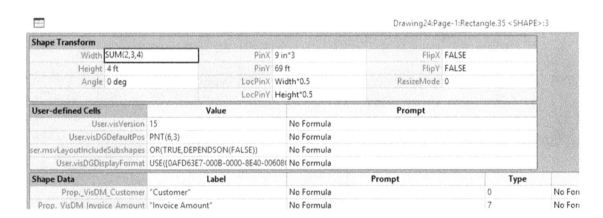

Step 7: Include the appropriate arguments for the function. You can enter them by typing them, by selecting appropriate ShapeSheet cells, or by using the NAME function to select ShapeSheet cells.

Step 8: Press the ENTER key to accept changes to the formula. Or you can press ESC to cancel the changes.

New Electrical Solution with IEEE (Institute of Electrical and Electronics Engineers) Compliance

There is improved support for industry standard templates, including engineering diagrams, operations, and quality industries. There are refreshed templates and thousands of shapes that meet industry standards including Unified Modeling Language (UML) 2.4, Business Process Model and Notation (BPMN) 2.0 and Institute of Electrical and Electronics Engineers (IEEE) compliance.

New Starter Diagrams and Contextual Tips and Tricks

Visio 2016 offers new starter diagrams that support industry standard workflows. We will start by looking at some predefined starter diagrams provided in the application.

Step 1: Click Categories.

Suggested searches: Network   Software

FEATURED   **CATEGORIES**

Step 2: Click Flowchart.

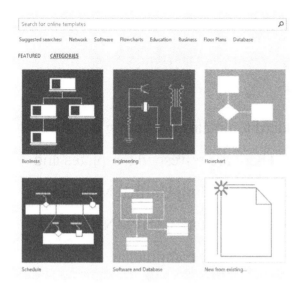

Step 3: single-click the Basic Flowchart thumbnail.

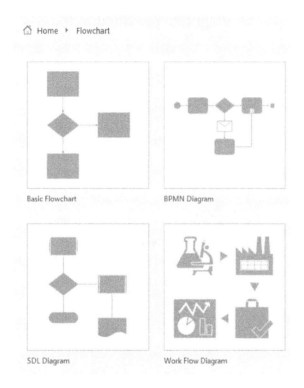

Step 4: Let's review what this dialog is all about.

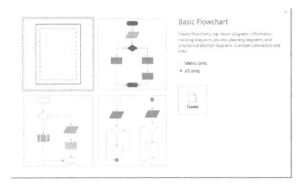

There are four templates. You can choose the blank template if you have some experience with Visio or choose one of the other three you do not have any experience

Step 5: Double-click one of the starter diagram thumbnails.

Starter diagrams in Visio 2016 can give you ideas and examples that you can customize by providing your own text and shapes. The starter diagrams have tips and tricks to help you work with the diagram.

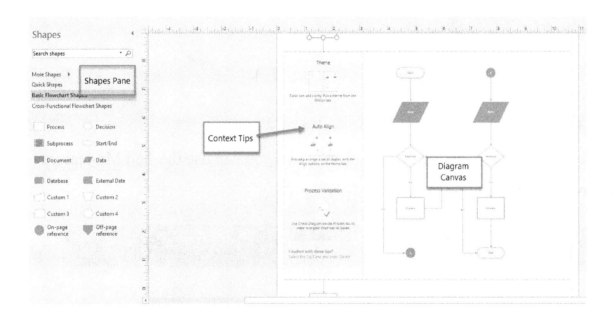

Starter Contextual Tips and Tricks

The starter diagrams have on screen contextual tips and tricks to help you get started with the diagram. To remove the contextual tips, select the tip and press Delete on the keyboard.

Tell Me Integration to Navigate Through Commands

Instead of searching the online help or in Visio, you can use the Tell Me feature to look for the solution you need.

Turn On/Off Page Auto Size for Visio 2007 and Earlier Files

A new diagram typically begins with a drawing page the size of a standard piece of printer paper. It is common for the diagram beyond the size of one printed sheet. Visio 2016 has dynamic page sizing capability that adjusts the size of the page as you draw. As you draw beyond the edge of the current page, Visio expands in that direction by one additional printed paper sheet.

The auto sizing behavior is controlled using the Auto Size button on the Design tab. When Auto Size is enabled, this reflects the orientation and size of the paper.

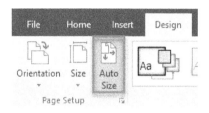

Disable the Auto Size option.

Step 1: Click the Design ribbon.

Step 2: Click the Auto Size button to toggle the Auto Size feature off or on.

Click the dialog launcher to open the Page Setup dialog where the legacy setting of "Same as printer paper size" has been replaced with the "Let Visio expand the page as needed." The "Size to fit drawing contents" option has also been moved to the Page size dropdown and renamed to "Fit to Drawing".

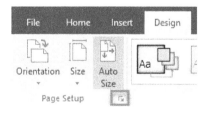

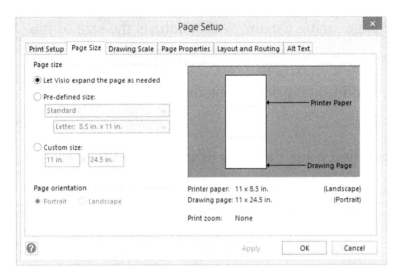

Turn On/Off High-Resolution Image Import

You can improve picture quality by turning off picture compression, but uncompressed pictures can result in very large file sizes.

Step 1: Click the picture once to open the Format tab of Picture Tools.

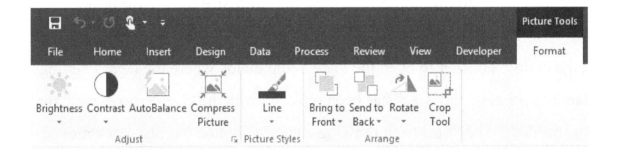

Step 2: Expand the Adjust menu.

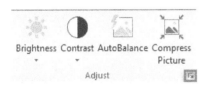

Step 2: Click the Compression tab.

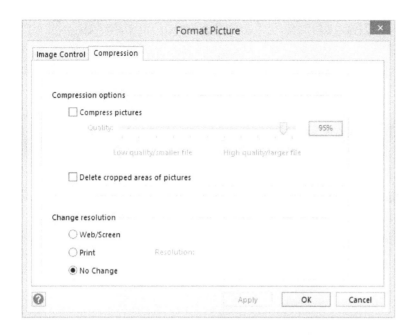

Step 3: Make the desired compression changes to the image.

Secure Diagrams with Microsoft File Protection Technologies

You can protect the diagram using Microsoft File Protection Technologies. You can find this feature on the File ribbon in Visio 2016.

Step 1: Click the File ribbon.

Step 2: Select protect Diagram on the Info menu.

Rapid Data Connectivity with Excel Tables (Ability to Swap Data graphics with one click)

Link without Dragging

Step 1: Click the Data tab.

Step 2: Click the External Data Window check box.

Step 3: Select the shape.

Step 4: Right-click a row of data and then click Link to Selected Shapes in the External Data Window.

Link many rows to many shapes

Step 1: Click the Data tab.

Step 2: Click Link Data on the Advanced Data Linking group.

Step 3: Follow the Automatic Link wizard instructions.

Step 4: In the automatically link row to shape if box:

Step 5: Select the column in the data source that contains unique values in the Data Column list.

Step 6: Select Shape Text in the Shape Field list or select a shape field previously created in the diagram.

Create Shapes from your Rows

Use this option if the drawing does not have shapes.

Step 1: Click a shape in the Shapes window.

Step 2: Drag a row or set of rows from the External Data window to a blank space in the drawing. An instance of the selected shape will appear for each row you drag to the canvas.

Step 3: Position the shape as nee.

Step 4: Click Change Shapes on the Home ribbon to change a shape to another shape.

Step 1: Click the Data ribbon.

Step 2: Click the Quick Import button in the External Data ribbon.

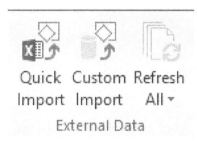

Step 3: Click Browse and select the Excel file containing the data you want to import.

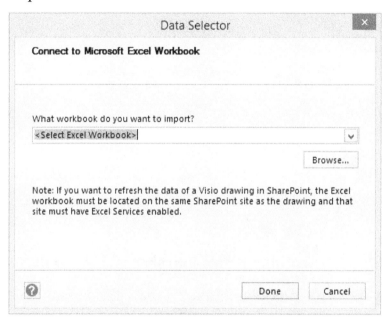

Step 4: Click Done to import the data file.

Custom Import of Data

Use the Custom Import to import other data types into the Visio diagram.

Step 1: Click Custom Import.

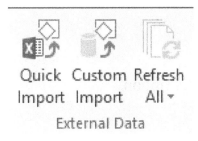

External Data

Step 2: Select the data source type.

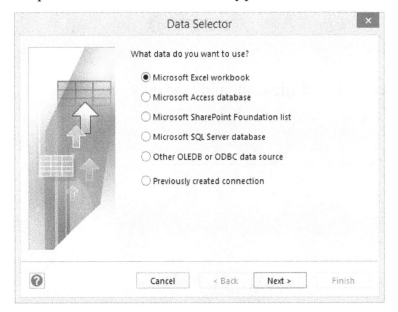

Step 3: Follow the prompts to import the data.

New Set of Data Graphics

Microsoft Visio Professional 2016 lets you import data into shapes. Visio will apply a data graphic to a shape after you import and link the data. Data graphics provide visualizations of the data for the shape. When you import data, Visio detects the type of data linked to the shape and then will apply an appropriate data graphic. You can insert a data bar, an icon, or a color assigned to a value.

- A data bar is good for showing a percentage, rating, progress, scores, and amounts.

- Icon sets can show percentages and progresses and can be useful for showing symbols for certain states or for showing Yes/No values.

- Text callout data graphics are like text labels and can be used to show data in the form of text or can be used in conjunction with other symbols related to currency.

- Color by value can be shown as a data graphic. A feature can be filled with a color based on the data. For example, in a network diagram, a computer might be colored red for off and green for on.

Access and Use Task Panes

To do this	Press
Move between selected submenu choices or move among some options in group of options in dialog box.	DOWN ARROW or UP ARROW
Move to a task pane from another pane in a program window	F6 or ALT + F6
When a menu is visible select the first command on the menu or submenu.	HOME
Float or anchor task panes	Press F6 repeatedly. Press ALT+SPACEBAR to open menu for task pane. Press the DOWN ARROW key to select Float Window command, and then press ENTER.
Open a shortcut menu	SHIFT+F10
Open selected menu, or perform action assigned to selected button	SPACEBAR or ENTER

To do this	Press
When a task pane is active, select next or previous option in task pane	TAB or SHIFT+TAB

Display and Use Windows

To do this	Press
Close active window	ALT+F4
Copy picture in selected window to Clipboard	ALT+PRINT SCREEN
Display shortcut menu for any window with an icon in title bar	ALT+SPACEBAR
Switch to next window	ALT+TAB
Open Reorder Pages dialog box	CTRL+ALT+P
Maximize selected window	CTRL+F10
Restore size of Visio program window after it has been maximized	CTRL+F5
Cycle focus through drawing pages, including any visible markup overlays	CTRL+PAGE DOWN
Cycle focus through pages in a drawing in reverse order	CTRL+PAGE UP
Cycle focus through open drawings in reverse order	CTRL+SHIFT+TAB

To do this	Press
	or CTRL+SHIFT+F6
Cycle focus through open drawings	CTRL+TAB or CTRL+F6
Copy a picture of screen to Clipboard	PRINT SCREEN
Open Page dialog box	SHIFT+F4

Edit Text

To do this	Press
Select all text in text block or select all master shapes in a stencil.	CTRL+A
Delete previous word	CTRL+BACKSPACE
Move down 1 paragraph (next paragraph)	CTRL+DOWN ARROW
Move to previous paragraph	CTRL+UP ARROW
Select the next paragraph	CTRL+SHIFT+DOWN ARROW
Select previous paragraph	CTRL+SHIFT+UP ARROW
Replace selected text with field height. If no text	CTRL+SHIFT+H

To do this	Press
selected, replace all text with field height for selected shape	
Select the next word	CTRL+SHIFT+RIGHT ARROW
Select previous word	CTRL+SHIFT+LEFT ARROW
Move to next line of text	DOWN ARROW
Move the previous line of text	UP ARROW
Move to the next character in a line of text	RIGHT ARROW
Select the next line	SHIFT+DOWN ARROW
Select next or previous character	SHIFT+RIGHT ARROW or SHIFT+LEFT ARROW
Select previous line	SHIFT+UP ARROW

To do this	Press
Open Home Tab ribbon	ALT+H
Turn subscript on or off	CTRL+=
Turn bold on or off	CTRL+B
Turn italic on or off	CTRL+I
Decrease the font size of the selected text.	CTRL+SHIFT+<
Turn superscript on or off	CTRL+SHIFT+=
Increase font size of selected text	CTRL+SHIFT+>
Turn all caps on or off	CTRL+SHIFT+A
Center text horizontally	CTRL+SHIFT+C
Turn double underline on or off	CTRL+SHIFT+D
Justify text horizontally	CTRL+SHIFT+J
Turn small caps on or off	CTRL+SHIFT+K
Align text left	CTRL+SHIFT+L
Center text vertically	CTRL+SHIFT+M

To do this	Press
Align text right	CTRL+SHIFT+R
Top-align text vertically	CTRL+SHIFT+T
Bottom-align text vertically	CTRL+SHIFT+V
Turn underline on or off	CTRL+U
Open Text dialog box	F11
Open Format Shape task pane	F3

Group, Rotate, and Flip Shapes

To do this	Press
Group selected shapes	CTRL+G or CTRL+SHIFT+G
Flip selected shape horizontally	CTRL+H
Vertically flip selected shape	CTRL+J
Rotate selected shape to left	CTRL+L
Rotate selected shape right	CTRL+R
Send selected shape to back	CTRL+SHIFT+B
Bring selected shape to front	CTRL+SHIFT+F

To do this	Press
Ungroup shapes in selected group	CTRL+SHIFT+U
Open Align Shapes dialog box for selected shape	F8

Work with Help

To do this	Press
Return to Visio Help home	ALT+HOME
Return to Previous Help topic	ALT+LEFT ARROW
Move to next Help topic	ALT+RIGHT ARROW
Switch between Help window and an active program	ALT+TAB
Perform action for selected item	ENTER
Perform action for selected Show All, Hide All, hidden text, or hyperlink	ENTER
Open Help window	F1
Scroll larger amounts up within currently displayed Help topic	PAGE UP
Scroll larger amounts down within currently displayed Help topic	PAGE DOWN

To do this	Press
Select previous item in Help window	SHIFT+TAB
Select next item in Help window	TAB
Select next hidden text or hyperlink, including Show All or Hide All at top of a topic	TAB
Scroll small amounts up or down, respectively, within currently displayed Help topic	UP ARROW or DOWN ARROW

Move Around in Application

To do this	Press
Activate hyperlink for shape or hyperlink on drawing that has focus	ENTER
Zoom in	ALT+F6
Zoom out	ALT+SHIFT+F6
Fit to window	CTRL+SHIFT+W
Exit full-screen view	ESC
Enter full-screen view	F5
Open next page in drawing or move to last master shape in a column of a stencil	PAGE DOWN

To do this	Press
Return to the previous page in the drawing or move to first master shape in a column of a stencil	PAGE UP
When a menu is visible, select the last common on the menu or submenu	END

Move Around in Text, Cells, or Dialog Boxes

To do this	Press
Move to end of text box	CTRL+END
Move to beginning of text box	CTRL+HOME
Move up 1 paragraph	CTRL+UP ARROW
Move 1 line down	DOWN ARROW
Move to end of line	END
Move to beginning of a line	HOME
Move 1 character to left	LEFT ARROW
Move 1 character to right	RIGHT ARROW
Move 1 line up	UP ARROW
Move 1 word left	CTRL+LEFT ARROW
Move 1 word to right or move to next word in a line	CTRL+RIGHT

To do this	Press
of text	ARROW
Nudge selected shape, toggle options in open drop-down list or between options in a group of options, or toggle between master shapes in a stencil	Arrow keys
Select previous hidden text or hyperlink, move from shape to shape on drawing page in reverse order, or move to previous option or option group	SHIFT+TAB
Move to previous character in a line of text	LEFT ARROW

Select Drawing Tools

To do this	Press
Select Pencil Tool	CTRL+4
Select Freeform Tool	CTRL+5
Select Line Tool	CTRL+6
Select Arc Tool	CTRL+7
Select Rectangle Tool	CTRL+8
Select Ellipse Tool	CTRL+9

To do this	Press
Select Pointer Tool	CTRL+1
Select text tool	CTRL+2
Select Connector tool	CTRL+3
Select connection point tool	CTRL+SHIFT+1
Select text box tool	CTRL+SHIFT+4
Switch Format Painter tool on or off	CTRL+SHIFT+P

Use Dialog Boxes

To do this	Press
Select an option, select or clear a check box	ALT+ Underlined letter option
Open selected drop-down list	ALT+DOWN ARROW
Switch to previous tab in dialog box	CTRL+SHIFT+TAB
Switch to next tab in dialog box	CTRL+TAB
Perform the action assigned to a default button in a dialog box.	ENTER

To do this	Press
Close a selected drop-down list; cancel a command and close a dialog box.	ESC
Open list if it is closed and move to option in list	First letter of an option in a drop-down list
Perform action assigned to selected button; select or clear the selected check box.	SPACEBAR
Move to the next option or option group	TAB
Move to file name box	ALT+N
Open selected file in the Open dialog box	ALT+O
Save current file in Save dialog box	ALT+S
Move to Save as type list in Save As dialog box	ALT+T
Move to file type list in Open dialog box	ALT+T
Perform action assigned to selected button	ENTER, SPACEBAR
Update file list	F5
Open Snap & Glue dialog box	ALT+F9

To do this	Press
Select or cancel selection 1 word to left	CTRL+SHIFT+LEFT ARROW
Select or cancel selection 1 word to right	CTRL+SHIFT+RIGHT ARROW
Move to end of entry	END
Move to beginning of an entry	HOME
Move 1 character to left or right	LEFT ARROW or RIGHT ARROW
Select from insertion point to end of entry	SHIFT+END
Select from insertion point to beginning of entry	SHIFT+HOME
Select or cancel selection 1 character to left	SHIFT+LEFT ARROW
Select or cancel selection 1 character to right	SHIFT+RIGHT ARROW

View Drawing Windows

To do this	Press
Display the open drawing windows so that you can see the title of every window.	ALT+F7 or CTRL+ALT+F7
Display open drawing windows tiled horizontally	CTRL+SHIFT+F7
Display open drawing windows tiled vertically	SHIFT+F7

Work with Shapes and Stencils

To do this	Press
Select Crop tool	CTRL+SHIFT+2
Select a shape that has focus. To select multiple shapes, Press TAB key to bring focus to first selected shape, press Enter. Hold down SHIFT key when pressing TAB key to bring focus to another shape. When focus rectangle is over the shape, press ENTER to add shape to selection. Repeat for each shape.	ENTER
Clear selection of or focus on a shape	ESC
Switch between text edit mode and shape selection mode on selected shape	F2
Nudge a selected shape 1 pixel at a time (Scroll lock must be turned off.)	SHIFT+Arrow keys

To do this	Press
Move from shape to shape on drawing page. A dotted rectangle indicates shape that has focus. Shapes that are protected against selection or on a locked layer cannot be moved.	TAB
Copy selected master shapes to Clipboard	CTRL+C
Insert selected master shapes into drawing	CTRL+ENTER
Paste contents of Clipboard to new stencil, when new stencil is opened for editing	CTRL+V
Move to last master shape in a stencil row	END
Cancel selection of master shapes in stencil	ESC
Move to first master shape in a row of a stencil	HOME
Select or cancel selection of a master shape that has focus	SHIFT+ENTER
Cut selected master shape from custom stencil and put it on Clipboard	CTRL+X
Delete selected master shape	DELETE
Rename selected master shape	F2